Lecture Notes in Earth Sciences 109

Max Wisshak

High-Latitude Bioerosion: The Kosterfjord Experiment

With 48 Figures, 3 in colour

Dr. Max Wisshak
Institute of Palaeontology
Loewenichstrasse 28
91054 Erlangen
Germany

E-mail: wisshak@pal.uni-erlangen.de

ISSN 0930-0317
ISBN-10 3-540-36848-5 Springer-Verlag Berlin Heidelberg New York
ISBN-13 3-540-36848-9 Springer-Verlag Berlin Heidelberg New York

Library of Congress Control Number: 2006930101

Springer is a part of Springer Science+Business Media
springer.com

Cover design: E. Kirchner, Heidelberg
Production: Almas Schimmel
Typesetting: camera-ready by author
Printing: Krips bv, Meppel
Binding: Stürtz AG, Würzburg

Printed on acid-free paper 30/3141/as 5 4 3 2 1 0

Preface

The study of bioerosion is situated at the interface between biology (the involved organisms), palaeontology (the borings they leave) and geology (palaeoecology, taphonomy, carbonate degradation and sediment production). Thus, the most promising approach to study bioerosion processes is an interdisciplinary one as well – a principle that has been acknowledged in the design of the Kosterfjord experiment, the first quantitative bioerosion experiment in a non-tropical setting. The targets of this experimental investigation were taxonomical and ecological aspects of bioerosion and their (palaeo)ecological implications alongside quantitative aspects of the carbonate cycling in the cold-temperate setting in the northeastern Skagerrak (SW Sweden).

The principal aims of the present volume are twofold: Firstly, I would like to give the reader a reference in hand that reviews the current state-of-the-art on cold-temperate to polar bioerosion processes and experimental bioerosion studies in general, and secondly, its immediate goal is to present the compiled outcome of the Kosterfjord experiment.

This contribution was worked out in the course of my PhD study and acts as doctoral thesis (dissertation) at the Friedrich-Alexander-University Erlangen-Nuremberg. In this context I would like to express my sincere thanks to André Freiwald, Richard Höfling and Joachim Reitner for furnishing the expert opinions. Partly incorporated in the present volume is the previously published outcome on the Kosterfjord experiment (Wisshak et al. 2005a and b; Wisshak & Rüggeberg 2006; Wisshak & Porter in press), and those references are consequently not explicitly cited, except for when referring to aspects omitted in this compilation. For these publications and the present volume, Tomas Lundälv contributed valuable environmental data, Marcos Gektidis carried out the biological identification and semi-quantitative analysis of the microendolithic organisms, and André Freiwald provided the discussion on palaeoenvironmental implications concerning cold-water coral occurrences. In the same context, I thank David Porter and Andres Rüggeberg for contributing their expertise on chytrid fungi and foraminiferans, respectively. Moreover, I would like to seize the opportunity to emphasise my gratitude to the various referees of the above-mentioned articles as there are Elisabeth Alve, Richard Bromley, Ingrid Glaub, Stjepko Golubic, Gerhard Schmiedl, Klaus Vogel and one anonymous colleague.

The Kosterfjord experiment was a team effort and I am deeply indebted to my Swedish project partner Tomas Lundälv, to Marcos Gektidis, and particularly to my mentor André Freiwald for their generous cooperation.

This study benefited from many fruitful discussions with colleagues from the bioerosion research community exemplified here by Ingrid Glaub, Stjepko Golubic and Richard Bromley. I gratefully acknowledge the Tjärnö Marine Biological Laboratory staff for their logistic and scientific support, especially Hans G. Hannsson, Lisbeth Jonsson, Bertil Rex and Lillemor Svärdh. Furthermore, my dear colleagues at the Institute of Palaeontology in Erlangen, specifically Tim Beck, Lydia Beuck, Sonja-B. Löffler, Matthias López Correa, Axel Munnecke and Jürgen Titschack induced much inspiration during countless discussions, and Birgit Leipner-Mata, Marie-Luise Neufert and Christian Schulbert provided valuable technical support. Also, I would like to express my gratitude towards André Freiwald, Marcos Gektidis, Leif Tapanila, and especially Sonja-B. Löffler and Stefanie Wisshak for their commitment in proofreading earlier drafts of this manuscript.

Sincere thanks are furthermore expressed towards the LNES editor Joachim Reitner and the publishing house Springer for their support in realising this contribution.

At last, elaborate science in these days is impossible without proper funding, which in the case of the Kosterfjord experiment was provided by the Deutsche Forschungsgemeinschaft (DFG-FR 1134/5-1-3). Furthermore, this research was supported in parts by the HERMES project, EC contract no GOCE-CT-2005-511234, funded by the European Commission's Sixth Framework Programme under the priority 'Sustainable Development, Global Change and Ecosystems'.

I would be most delighted if the present volume stimulates further neontological or palaeontological studies in the wide field of high-latitude bioerosion and furthers the understanding of the ecological complexity of carbonate buildup and degradation.

Erlangen, November 2005

Max Wisshak

Contents

List of specific abbreviations and units

BP = before present
BS = endolith in situ in translucent bivalve shell
CCD = carbonate compensation depth
CT = cold-temperate
CTD = conductivity, temperature, depth
FISH = fluorescence in situ hybridisation
GBR = Great Barrier Reef
GIS = geo information system
I = isolated microendolith
ICBN = International Code of Botanical Nomenclature
ICZN = International Code of Zoological Nomenclature
ILM = incipient light microscopy
IPAL = Institute of Palaeontology, Erlangen, Germany
IS = endolith in situ in Iceland spar
L = lagoon
M = arithmetic mean value
P = polar
PAR = photosynthetically active radiation
PVC = polyvinyl chloride
PP = polypropylene
ROV = remote operated vehicle
SD = standard deviation
SEM = scanning electron microscopy
SST = sea surface temperature
T = tropical
TLM = transmission light microscope
TMBL = Tjärnö Marine Biological Laboratory, Sweden
WT = warm-temperate

psu = practical salinity units
ntu = nephelometric turbidity units
Ω = carbonate saturation state of sea-water

1 Introduction

1.1 The process of bioerosion

Bioerosion is the major force driving the degradation of carbonate skeletal material and rocky limestone coasts in all marine and some freshwater environments in concert with physicochemical dissolution and mechanical abrasion. Even though the ability of various hardground dwelling biota to degrade lithic substrates is known for a long time, it was not before the late 1960's when the term 'bioerosion' was introduced to the literature by Neumann (1966) as "the removal of consolidated mineral or lithic substrates by the direct action of organisms". In the present volume, bioerosion exclusively refers to the degradation of marine carbonates, which represent by far the most relevant and best studied substrate. Biotic boring activity also occurs in wood, bone, siliciclastic and even crystalline rock, and is thereby not restricted to marine environments.

Soon after Neumann's (1966) pioneering experimental work, the ecological and geological significance of biological carbonate degradation was realised and bioerosion came into a wider focus of biologists and actuopalaeontologists, now concentrating on community structures and processes. The study of microendoliths was catalysed by the development of a suitable preparation and visualisation method by Golubic et al. (1970) – SEM analysis in combination with the cast embedding technique. During the following decades, a wealth of studies has been carried out on marine bioerosion in tropical and warm-temperate seas supporting an extensive literature. Studies in high-latitude, cold-temperate to polar settings on the other hand remain relatively sparse (Sect. 1.2). This is particularly obvious with respect to experimental approaches, which were previously almost exclusively carried out in tropical seas (Sect. 1.3). Settlement experiments investigating calcareous epizoans were also situated in high-latitude settings (Sect. 1.4).

A wide range of mechanical and/or chemical boring, scraping, biting, crushing or gnawing organisms are known to break down calcareous substrates. Those bioerosive agents comprise grazers (gastropods, chitons, echinoids etc.), macroborers (such as sponges, bryozoans and worms with traces >100 μm in diameter) and microborers (mainly bacteria, fungi and algae with traces <100 μm in diameter). In fact, representatives of about a dozen phyla and groups of at least two out of the three domains (Eucarya and Bacteria) are present – not only in tropical and warm-temperate seas but also in the cold-temperate and polar biogeographic regions (Sects. 1.5 and 1.6).

The traces left by bioerosion agents on and within hard substrates yield a fossilisation potential far superior of their producers, and may become part of the fossil record as trace fossils that can be classified as ichnotaxa (Sect. 1.7). Microboring traces especially yield an extensive fossil record reaching as far back as the Proterozoic (Bromley 2004; Glaub & Vogel 2004 and references therein). From an evolutionary point of view, microborers are remarkably conservative organisms (in morphological as well as in palaeobiological respects), making them well-suited palaeoenvironmental indicators for judging palaeobathymetry, palaeosalinity or palaeotemperature (Sect. 1.8).

By representing the chief carbonate degrading process, bioerosion plays an especially relevant role in the global carbon(ate) cycling as an opposing player of carbonate precipitation (Sect. 1.9). Therefore, the relative importance of bioerosion relative to physicochemical dissolution is a subject of ongoing debate and investigation (Sect. 1.10).

For general introductions to different aspects of marine bioerosion, the reader is referred to the review papers by Bromley (1970), Golubic et al. (1975) and Warme (1975). A review on the various ichnotaxa and their stratigraphic record can be found in the 'Trace Fossil Treatise' volume (Häntzschel 1975) and in the recent compilations of macroboring trace fossils by Bromley (2004) and microborings by Glaub & Vogel (2004). A recent review on marine hardground communities (borers and encrusters) was provided by Taylor & Wilson (2003). Extensive bibliographic sources are Clapp & Kenk (1963), Radtke et al. (1997) and the continuously growing online bibliography maintained by Wilson (2005).

The two-year bioerosion experiment reported in this volume was designed and launched in September 2002 in close cooperation with the Tjärnö Marine Biological Laboratory (TMBL) in the cold-temperate setting of the northern Kosterfjord area off SW Sweden. This interdisciplinary study tackled various ecological and taxonomical aspects of bioerosion and carbonate accretion in qualitative as well as quantitative respects (Sect. 1.10).

1.2 High-latitude bioerosion studies

In the following account, a brief overview on previous work on bioerosion in high-latitude settings is given. Hereby, the term 'high-latitude' refers to the cold-temperate and polar biogeographic regions. The boundaries between these realms (Fig. 1) are conventionally delineated by mean sea surface temperatures (SST) with the 15°C summer and 10°C winter isotherm marking the boundary between the warm- and cold-temperate regions and the 10°C summer and 0°C winter isotherm drawing the boundary between the cold-temperate and polar seas, respectively (Lüning 1985). Most studies were undertaken in the North Atlantic and north-eastern Pacific, while references from the southern hemisphere are as yet very scarce.

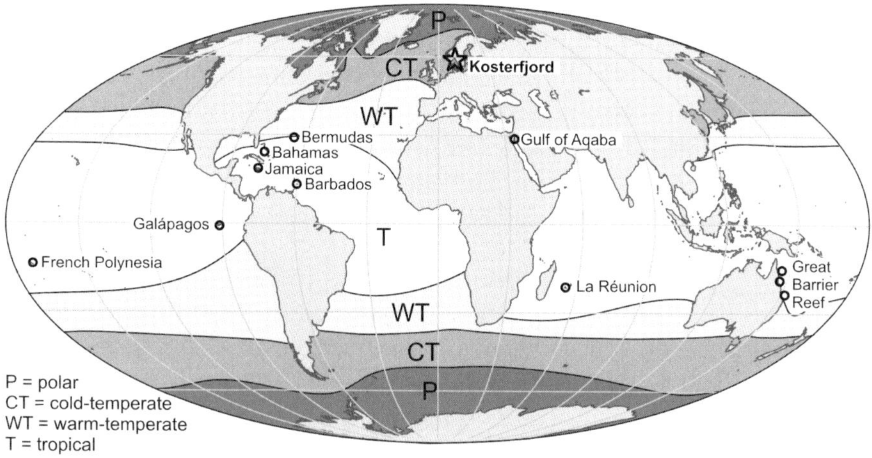

Fig. 1 The marine biogeographic regions (modified after Lüning 1985) and the study sites of previous quantitative experimental bioerosion studies

Macrobioerosion

An early review on marine macrobioerosion in the cold-temperate British waters was given by Jehu (1918). Boring predation was reviewed by Carter (1968) in a case study of molluscs from temperate and polar settings. Both, macroborers and grazers were briefly reviewed by Wilson (1982).

Boekschoten (1966) presented an extensive and important study on shell boring organisms (macroborer and microborer) recorded in shells from the Dutch coast. The temperature related boring progress of polychaetes and sponges was observed with the aid of time lapse radiography in bivalve shells from Newfoundland by Evans (1969). The same agents were studied in detail by Hoeksema (1983) along the Dutch coast. On the Scott Shelf off W Canada, boring sponges were investigated by Young & Nelson (1985) who assessed the contribution of sponge chips to fine-grained carbonate components and drew attention to the importance of this bioerosion agent also in temperate seas with low carbonate production and sedimentation rates. Farrow & Clokie (1979) and Akpan (1984) presented a detailed investigation of grazing limpets, chitons and echinoids, their differing feeding habits and linked bathymetric distribution, and their importance in carbonate mud production. Richard Bromley and different co-workers studied a number of bioeroders and related ichnotaxa, also referring to Recent examples from high-latitude areas, as for instance brachiopods (Bromley & Surlyk 1973), echinoids (Bromley 1975), gastropods (Bromley 1981), boring bivalves (Kelly & Bromley 1984), anomiid bivalves (Bromley & Martinell 1991), octopi (Bromley 1993) and presumed foraminiferans (Bromley et al. in press). From a setting situated in 70° northern latitude (Vardø Island, N Norway), Bromley & Hanken (1981) presented the northernmost analysis of a boring community to date. Also north of the Arctic Circle but on the other side of the Atlantic in the eastern Canadian Arctic, Aitken & Risk (1988) studied boring traces – primarily gastropod and polychaete borings – recorded in Recent to subrecent mollusc shells, calcareous algae and limestone clasts.

Microbioerosion

The importance of marine fungi (Phycomycetes) in the degradation of calcareous shells was recognised by Zebrowski (1937), working on Recent shell material sampled in the cold-temperate southern Australian waters. The topic was picked up by Höhnk (1969), studying shell material from the North Sea. Kohlmeyer (1969) gave a first review on boring fungi and Cavaliere & Alberte (1970) presented additional observations from shell material collected along the coasts of Iceland.

Based on the pioneering taxonomic work on endolithic algae and cyanobacteria by Bornet & Flahault (1888, 1889), Kylin (1935) furthered our knowledge on these bioerosion agents with his investigation at the Swedish west coast. Kornmann (1959, 1960, 1961) and Kornmann & Sahling (1980) made significant progress in the understanding of various endolithic chlorophytes during laboratory studies and observations in the vicinity of Helgoland. Subsequently, Wilkinson & Burrows (1970, 1972) and Wilkinson (1974, 1975) picked up the topic with detailed studies in Britain and E Canada. Ruth Nielsen deepened the knowledge on endolithic algae from high-latitudes in both hemispheres with numerous studies (e.g., Nielsen 1972, 1980, 1987; Nielsen & Correa 1987).

Endolithic rhodophytes from high-latitudes were reported by Conway & Cole (1977) and subsequently, Clokie and co-workers (1979, 1980, 1981) recognised the importance of rhodophyte *Conchocelis*-stages for the delineation of the photic limit in their studies on the Rockall Plateau (off Ireland) and the Firth of Clyde (Scotland).

Stimulated by the work on the influence of microboring activity on the genesis of micrite envelopes in tropical settings by Bathurst (1966, 1971), Alexandersson (1972, 1974a and b, 1976, 1977, 1978, 1979) presented an array of comparative studies from cold-temperate settings and especially from the Skagerrak and Kattegat. He specifically observed bioerosion as well as maceration, and related the latter process to apparently carbonate undersaturated waters of cool- to cold-water bodies (see Sect. 1.10). This contribution highly influenced the understanding of cool-water carbonate diagenesis and is still often referred to as well as disputed on. In further contributions on the same topic, Gunatilaka (1976) investigated algal borings and micritisation of skeletal sands off W Ireland, Scoffin et al. (1980a) observed the diagenetic processes of carbonate grains in sub-photic depths at the Rockall Bank (off NW Ireland), and Cutler & Flessa (1995) compared intertidal shell preservation from the Skagerrak with those from the Gulf of California. A recent comprehensive compilation on taphonomic and diagenetic processes on temperate versus tropical seas was given by Smith & Nelson (2003).

The most extensive studies of high-latitude macro- and microbioerosion were undertaken by G. Farrow, E. Akpan and co-workers during the 1980's in various settings along the coast and shelf of Scotland: Akpan (1984) gave a first brief report on microendolithic algae and fungi. Akpan & Farrow (1984a and b, 1985) and Akpan (1986) put further efforts in characterising microendoliths and their communities as well as their bathymetric distribution. Especially relevant in this respect is their extensive study on shell boring algae and their bathymetric zonation (1984a) which paved the road for the palaeobathymetric scheme established later by Glaub (1994, 1999; see Sect. 1.8). In addition, Farrow et al. (1984) evaluated the contribution and importance of bioerosion to bioclastic carbonate sedimentation. As a final outcome, Farrow & Fyfe (1988) presented a comprehensive review on high-latitude bioerosion and carbonate mud production.

From the cold-temperate Pacific, a first extensive investigation of microendoliths was given by Hernderson & Styan (1982), who studied shell material recovered off British Columbia. They were especially focusing on the co-occurrence of algal and fungal traces and their potential trophic relationship, finding that fungi possibly utilise algae as an energy source. Young & Nelson (1988) gave additional detailed insight into the micro- and macroboring communities studied from the Scott Shelf off W Canada. The study was complemented by laboratory tumbling experiments which well demonstrated the catalysing effect of bioerosion for mechanical breakdown and abrasion.

Further miscellaneous work on microbioerosion was done by Boekschoten (1966) on shell boring organisms recorded in shell material from the Dutch coast, by Gaspard (1989) on microendolithic biodegradation of Recent brachiopod shells from Scottish and New Caledonian waters, and by Cerrano et al. (2001) providing a brief report of microbioerosion in Antarctic seas. Glaub et al. (2002) investigated the microendolithic inventory of various North Atlantic settings (e.g., N Scotland and N Norway) with special reference to the extension of the photic zonation and the applicability of the bathymetric index ichnocoenoses established by Glaub (1994, 1999). Microendoliths and their bathymetric distribution were furthermore tackled from the Tromsø area (N Norway) by Schmidt & Freiwald (1993).

Freiwald (1995, 1998) drew attention to carbonate dissolution induced by the degradation of skeletal binding organic matter by bacteria (maceration), studied in foraminiferal tests and echinoid spines from the sediment water interface of a shallow coastal platform in northern Norway. Freiwald & Wilson (1998) studied aphotic bioerosive processes such as biofilm and fungal infestation as well as the activity of boring sponges controlling the breakdown of the cold-water coral *Lophelia pertusa*. A detailed analysis of the boring traces recorded in the latter scleractinian coral from the Sula Ridge off Norway was given by Krutschinna (1997) and Vogel et al. (1999); from the Porcupine Seabight off Ireland by Beuck & Freiwald (2005). The boring foraminiferan *Hyrrokkin sarcophaga* parasitising on this coral was studied by Freiwald & Schönfeld (1996). Bioerosion by sponges and fungi in another cold-water coral – *Desmophyllum cristagalli* – was reported by Boerboom and co-workers (1998) from a NE Atlantic seamount and by Försterra et al. (2005) from a Chilean fjord site.

Bioerosive agents and processes shaping the intertidal zone of coastal limestone were evaluated in detail by Trudgill and co-workers in western Ireland (Trudgill 1987; Trudgill & Crabtree 1987; Trudgill et al. 1987). Studies on the same topic, including quantitative investigations were carried out by Kelletat (1986) in Scotland, Ireland (Kelletat 1988), and from the opposite side of the globe at the coasts of New Zealand (Kelletat 1987). By comparing his results from high-latitude settings with his previous studies at low-latitude sites in the Mediterranean Sea and Australia, and the data available from the literature, he found a raised littoral bioerosion activity in cool and cold environments (Kelletat 1988).

1.3 Previous experimental bioerosion studies

A number of simple to highly sophisticated bioerosion experiments were carried out since the 1950's, monitoring various environmental parameters that influence bioerosion, and investigating boring community development as a function of time, eutrophication, specific reef environments or bathymetry. Some of these studies followed a quantitative approach by assessing specific bioerosion rates and partly also carbonate accretion rates (Table 1; Fig. 1).

Pioneer studies

First studies on the bioerosion rates of reef limestone and beachrock were undertaken in the 1950's to 60's – often by direct measurements of inter- and supratidal surface retreat (see Neumann 1966 and Trudgill 1983, for brief reviews and references therein). During this pioneer phase of quantitative carbonate erosion studies, the importance of biological processes and especially bioerosion as opposed to chemical dissolution and physical erosion was recognised. The study by Neumann (1966) not only introduced and defined the term 'bioerosion', but was also among the first to determine bioerosion rates through a series of field and laboratory experiments carried out on the Bermudas.

In the following decade, first efforts were undertaken to quantify the erosion capability of various specific macroboring and grazing agents such as fish (e.g., Brock 1979; Scoffin et al. 1980b), echinoderms (e.g., McLean 1967; Hunter 1977; Scoffin et al. 1980b), molluscs (e.g., McLean 1967; Trudgill 1976) and sponges (e.g., Neumann 1966; Hein & Risk 1975; Rützler 1975; Moore & Shedd 1977; Bromley 1978).

During the 1970's, the detailed study of microendoliths and their communities was catalysed by the development of the cast embedding technique by Golubic et al. (1970). In the following years, a number of important studies were carried out, some of which following an experimental approach by investigating planted mollusc shells and/or Iceland spar. Relevant studies in this context are the 3-year experiment carried out on the Puerto Rican shelf by Golubic (1969), the experiments at a reef on the St. Croix shelf (West Indies) by Perkins & Tsentas (1976), the studies of algal infestation and micritisation by Kobluk & Risk (1977a and b), and the only experimental study investigating microendoliths below the sediment/ water interface by May & Perkins (1979).

Table 1 A compilation of previous quantitative experimental bioerosion studies with the exposure periods, substrates and quantification methods used, and the assessed carbonate accretion, micro-, macro-, total and net bioerosion rates

Study site	Setting and water depth [m]	Substrate	Exposure time [months]	Carbonate accretion [g /m²/y]	Bioerosion grazer [g /m²/y]
Carribean Sea (USA)					
Lee Stocking Island, Bahamas	Reef transect down to off reef slope; 275 m	Micrite blocks	3 6 12 24		
Black Rock, Bahamas	Intertidal to 477 m	Micrite blocks	12		
French Polynesia (France)					
Tiahura Reef Moorea	Barrier reef flat	*Porites* blocks	2 6 12 24		-370 -960 -1,740 -2,330
Moorea, Tahiti, Tikehau and Takapoto Island	Fringing reefs and atoll lagoons; 1-2 m	*Porites* blocks	6	60 to 940	-300 to -10,380
Moorea, Tahiti, Tikehau and Takapoto Island	Fringing reefs and atoll lagoons; 1-2 m	*Porites* blocks	24	180 to 1,130	-330 to -6,870
Moorea, Tahiti, Tikehau and Takapoto Island	Fringing reefs and atoll lagoons; 1-2 m	*Porites* blocks	60	30 to 230	-1,270 to -2,490
Great Barrier Reef (Australia)					
Davies Reef	Lagoon 5 m	Fresh molluscan sand 1-4 mm	12		
Wreck, One Tree and Llewellyn Island	Reef slope, reef flat and lagoon sites; 0.5 to 10 m	*Porites* blocks	6 12 18 24	90 to 970 0 to 630 0 to 1,040 0 to 2,150	
Lizard Island	Reef flat 1.5 m, reef slope; 10 m, lagoon 7 m	*Porites* blocks	24+30 36+42		-20 to -760 -10 to -1,510

[A] Referring to m² lagoonal seafloor and not substrate area

Bioerosion macroborer [g /m²/y]	Bioerosion microborer [g /m²/y]	Bioerosion total [g /m²/y]	Accretion - Bioerosion [g /m²/y]	Method	Source
	~0 to -390			Image analysis of SEM images from epoxy resin casts	Vogel et al. 1996, 2000
	~0 to -520				
	~0 to -240				
	~0 to -210				
	0 to -259			Gravimetrical	Hoskin et al. 1986
-2	-570	-940		Image analysis	Peyrot-Clausade et al. 1995a; Chazottes et al. 1995
-3	-200	-1,160			
-22	-140	-1,900			
-90	-200	-2,620			
-20 to -260			+810 to -10,450	Volumetrical + point-counting of sections	Peyrot-Clausade et al. 1995b
-20 to -140			+630 to -6,770	Image analysis of sections	Pari et al. 1998
-20 to -1,040			-1,610 to -3,020	Image analysis of sections	Pari et al. 2002
	>-350[A]			Gravimetrical + point-counting	Tudhope & Risk 1985
		-40 to -610		Point-counting of sections + gravimetrical	Kiene 1988
		-250 to -1,710			
		0 to -5,910			
		-110 to -9,110			
-180 to -500		-220 to -1,260		Point-counting of sections	Kiene 1985
-400 to -1,200		-410 to -2,710			

Table 1 continued

Study site	Setting and water depth [m]	Substrate	Exposure time [months]	Carbonate accretion [g /m²/y]	Bioerosion grazer [g /m²/y]
Great Barrier Reef (Australia) continued					
Lizard Island	Various reef environments from 1 to 20 m	*Porites* blocks	24 36 60		-10 to -190 -10 to -120 -10 to -160
Lizard Island	Various reef environments from 1 to 20 m	*Porites* blocks	86-111		-350 to -2,460
One Tree Island	Various patch reef sites	*Tridacna* blocks	5		
One Tree Island	Various patch reef sites	*Porites* blocks	26		-300 to -1,090
Northern Great Barrier Reef	200 km cross-shelf transect (6 sites); 7-10 m	*Porites* blocks	12		-280 to -2,800
Northern Great Barrier Reef	200 km cross-shelf transect (6 sites); 7-10 m	*Porites* blocks	24 48	7 to 102[B] 27 to 224[B]	-39 to -1,700 -590 to -7,415
Indian Ocean					
Réunion Island	Fringing reef, three transects	*Porites* blocks	12		-1,510 to -4,310
Red Sea (Jordan)					
Aqaba, Gulf of Aqaba	Fringing reef, 5, 15, 25, 40 m	*Porites* blocks	12 24	16 to 236 37 to 196	-35 to -492 -42 to -857
Galápagos (Equador)					
Champion Island	Fringing reef, 5 to 13 m	*Porites* and micrite blocks	15 15	 600 to 1,000	-22,800 -3,500

[B] Coralline algae only

Bioerosion macroborer [g /m²/y]	Bioerosion microborer [g /m²/y]	Bioerosion total [g /m²/y]	Accretion - Bioerosion [g /m²/y]	Method	Source
-20 to -80 -50 to -200 -70 to -350		-150 to -2,000 -250 to -1,450 -200 to -1,900		Point-counting of sections	Kiene & Hutchings 1994b
-60 to -240		-410 to -2,520		Point-counting of sections	Kiene & Hutchings 1994a
	-8 to -43			SEM Image analysis of epoxy resin casts	Kiene 1997
-3 to -26				Image analysis of sections	Kiene 1997
-10 to -90	-120 to -1,340	-460 to -3,600		Image analysis of sections and SEM images	Tribollet et al. 2002
-39 to -283 -154 to -1,075			-180 to -1,748 -1,090 to -7,846	Image analysis of sections	Osorno et al. 2005
-6 to -47	-24 to -69	-1,560 to -4,370		Image analysis	Chazottes 1996; Chazottes et al. 2002
		-122 to -655 -238 to -1,241	+114 to -586 -83 to -2,298	Image analysis + gravimetrical	Hassan 1998
-2,600 -600		-25,400 -4,100		Image analysis of sections	Reaka-Kudla et al. 1996

French Polynesia

An array of experimental studies were conducted by the French bioerosion research community (V. Chazottes, T. Le Campion-Alsumard, M. Peyrot-Clausade, N. Pari and others) during the late 1980's and early 90's at various sites in French Polynesia.

A first experiment was launched 1986 at the Tiahura Reef flat (Moorea), where *Porites* blocks were planted and a set of bioerosion rates was determined by Peyrot-Clausade et al. (1995a) and Chazottes et al. (1995). In addition, a series of in situ and laboratory experiments was carried out, investigating various aspects of the reef's carbonate budget (Le Campion-Alsumard et al. 1993; Peyrot-Clausade et al. 1995a).

The main experiment was launched in 1990 at various sites on reefs fringing Tahiti and Moorea and in lagoonal settings of Tikehau and Takapoto. These sites provided a well suited ground to study the variation of bioerosion rates and patterns in pristine atoll lagoons with little river runoff compared to fringing reefs surrounding high volcanic islands with varying levels of eutrophication due to high river runoff during wet seasons and high anthropogenic pollution impact. First results were published by Peyrot-Clausade et al. (1995a and b) focusing on the statistical population analysis of the macroborers and grazers, and determining further bioerosion and carbonate accretion rates. They recognised the inverse relationship of macroborer density and grazing pressure, and the positive effect of eutrophication on grazing activity. The development of epilithic and endolithic algae populations on the test blocks and background samples were evaluated with respect to hydrographic data by Le Bris et al. (1998).

The experiment was extended to 60 months total exposure and the final reports (Pari et al. 1998, 2002) documented detailed bioerosion rates and an extensive statistical analysis of the variability between and within sites and the development with time. Their data showed, that epilithic algae growth was propelled by eutrophication at the fringing reefs resulting in higher accretion rates whereas endolithic algae were more abundant at the pristine lagoonal sites. They concluded that these site and time variations of bioerosion rates are characteristic for reefs in general. Additionally, they pointed out the importance of maintaining a good water quality in order to keep bioerosion rates at a lower level as an important prerequisite for the recovery and protection of tropical reefs.

The French Polynesian experiments were rounded out by a number of qualitative studies on microendoliths undertaken on background sample material (Peyrot-Clausade et al. 1992; Le Campion-Alsumard et al. 1995a and b; Tribollet & Payri 2001; Hutchings & Peyrot-Clausade 2002).

Indian Ocean

Additional experiments were launched by the French working group at Réunion Island. The results were compared to their French Polynesian findings: At three transects of a fringing reef along Réunion's west coast, *Porites* blocks were deployed to study the influence of eutrophication on agents and rates of bioerosion (Chazottes 1996; Chazottes et al. 2002). They found significant differences between sites as a function of nutrient availability, determining the degree of epilithic substrate coverage and consequently enhanced grazing attraction. These results were confirmed by a study of internal bioerosion agents and rates in another scleractinian (*Acropora*) at the same study site by Zubia & Peyrot-Clausade (2001). The assessed bioerosion rates were incorporated in a calculation of the carbonate budget in a comparative study with the French Polynesian sites (Peyrot-Clausade et al. 1999).

Great Barrier Reef

Another series of long-term experiments designed to assess the distribution and dynamics of bioerosion processes and rates, chiefly conducted by W.E. Kiene and P.A. Hutchings targeted various reef settings at the Great Barrier Reef (GBR) in Australia. The main site of these experiments was Lizard Island, where *Porites* blocks were planted in various reef settings ranging from the shallow leeward lagoon to 20 m water depth at the windward slope, between 1980 and 1983. First results of the experiments and especially on the recruitment patterns of boring polychaetes were published by Hutchings & Murray (1982), Davies & Hutchings (1983) and Hutchings (1984, 1985). First bioerosion rates were assessed by Kiene (1985). The authors revisited the sites repeatedly and outlined the further development of the boring community and bioerosion rates extending the overall exposure time to 9 years (Hutchings et al. 1992; Kiene & Hutchings 1994a and b). At Lizard Island, grazers (chiefly scarids) were the dominating agents of bioerosion followed by boring sponges. The latter as well as boring polychaetes and molluscs exhibited a highly variable distribution in time and space, which can be regarded as a general pattern for macroborer dispersal. Additional important results of this experimental study were the interaction and interference of grazing and boring activity and their implication for the interpretation of the maturity of boring communities and bioerosion rates.

Another 2-year experiment was conducted by Kiene (1988) on three islands in the southern GBR, investigating bioerosion and accretion (rates and agents) on reefs in different stages of reef evolution. Kiene pointed out the significance of the balance between bioerosion and accretion on the one hand and grazing and boring on the other, determining the preservation potential of a reef framework through time. The latter was highest for the reef flat of mature reefs and lowest for the slope of immature settings. The study led to the development of a modified model for the various processes affecting a dead reef framework, with focus on the alteration rather than destructive influence of grazers and borers.

Yet another experiment targeted the response of bioerosion and specifically microbioerosion to nitrogen and phosphorous treatments as part of the ENCORE nutrient enrichment experiment at One Tree Island (Larkum & Steven 1994). The bioerosion experiments comprise a detailed qualitative study of the microendoliths in planted Iceland spar, micrite blocks and mollusc substrates as well as a quantitative assessment of bioerosion rates, following the experimental design of the earlier GBR studies (Kiene 1997; Vogel et al. 2000). They found no significant influence of the nutrient treatments on the rates of grazing, micro- and macroboring, but pointed out the possible requirement of higher doses and longer exposure times before eutrophication effects become detectable.

An independent 1-year experiment was carried out by Tudhope & Risk (1985) at Davies Reef, determining the rate of carbonate degradation by microborers in mobile lagoonal sediments instead of sessile calcareous substrates by employing molluscan sand-sized grains in experimental holders.

In the mid 1990's, the last experimental study was launched, comprising 6 stations along a 200 km transect across the northern GBR in order to evaluate the influence of terrigenous input and spatial variability of bioerosion (Tribollet et al. 2002). In this study, microbioerosion rates were evaluated in addition to the usual grazing and macroboring rates. They turned out to be of important quantitative relevance during early stages (1 year) of exposure, by far exceeding the rate of macrobioerosion. Hutchings et al. (2005) and Osorno et al. (2005) extended the experiment to a total of 4 years exposure and evaluated the various bioerosion agents and rates for instance with respect to eutrophication by local river runoff.

Red Sea

The basic experimental design developed by Kiene (1985) in the GBR experiments was recently also applied in the coral reef province of the Red Sea by Hassan (1998). Her main aim was to reveal variations in bioerosion and accretion along a bathymetric transect down to 40 m water depth across a fringing reef at Aqaba. In addition, Hassan compared the results of this setting with a hydrologically more exposed setting in the central Red Sea and with computer tomography of background coral samples.

Galápagos Islands

Reaka-Kudla and co-workers (1996) recorded bioerosion rates at a severely bioeroded coral reef at Champion Island during a 1-year experimental study. This study yielded results on differences between coral and micritic limestone substrates as well as the interplay of accretion and erosion.

Caribbean Sea

A remarkable study assessed the carbonate budget of a fringing reef on Barbados (Stearn & Scoffin 1977; Scoffin et al. 1980b). This extensive study of the various processes contributing to carbonate build-up and degradation comprises estimates on bioerosion rates of macroborers, parrot fish and echinoids, and additionally includes a detailed qualitative study of macroboring communities and grazers.

During the early 1980's, the Bahama carbonate factory was the target of an experimental study on carbonate production, transport and degradation (Hoskin et al. 1986). In the course of this study, bioerosion rates also were assessed applying in situ enclosure experiments and planted micrite blocks. More recently, the German working group led by K. Vogel also conducted bioerosion experiments on the Bahamas (Kiene et al. 1995; Vogel et al. 1996, 2000; Gektidis 1997a and b, 1999; Vogel 1997). The main goals of their studies at Lee Stocking Island were to assess factors influencing the distribution of microendoliths, and the calibration of the bathymetric index ichnocoenoses scheme established and applied by members of the same working group (e.g., Glaub 1994, 1999). The study included a quantitative assessment of microbioerosion rates at the various reef environments, covering an extended bathymetric transect down to 275 m water depth (Vogel et al. 2000). In addition, the distribution of microendoliths in different substrates and in various reef environments was investigated in detail, such as the experimental study of microbioerosion in ooids by Gektidis (1997a), revealing significant differences in boring communities in mobile versus fixed ooid grains.

Another 6-months experimental study aimed to carry out detailed cytological investigations of various endoliths and led to the discovery of endolithic thraustochytrid fungi in planted shell fragments in the shallow waters of Discovery Bay on Jamaica and additionally at coastal sites of Maine and Georgia (Porter & Lingle 1992).

Mediterranean Sea

During the late 1970's, Le Campion-Alsumard (1975, 1978, 1979) conducted a number of detailed experimental investigations (employing artificial carbonate substrates) of endolith colonisation along the French limestone coast in the Marseille region. These studies provided many qualitative observations on the morphology, taxonomy and ecology of endolithic chlorophytes and especially cyanobacteria. They furthermore gave some quantifications of percentage cover with exposure time and penetration densities per square centimetres (but no bioerosion rates) in the supra- to shallow-subtidal zones.

Extended long-term experiments

Only a small number of extended long-term experiments were carried out, some of which are still ongoing. Those studies are important since community succession of macroborers takes many years to reach equilibrium as opposed to microbioerosion communities, which – at least in tropical settings – develop mature ichnocoenoses in a much shorter period of time.

Scott et al. (1988) deployed concrete cinder blocks and coral limestone rubble for 13 years in 5 m water depth at Discovery Bay (Jamaica). There, polychaetes, the bivalve *Lithophaga* and the boring sponges *Cliona* and *Damiria* accounted for 4.5% erosion by volume of the carbonate substrates after 13 years.

A still ongoing experimental long-term bioerosion study was launched in the early 1980's by R. G. Bromley & U. Asgaard (Bromley et al. 1990; Bromley & Asgaard 1999) on the Island of Rhodes (Greece) where they deployed marble blocks at 3 to 17 m water depth. The study focuses on community successions of boring, grazing and encrusting organisms as a function of exposure time, water depth and different hydrodynamic settings. Bioerosion rates were as yet not determined but are intended to be assessed applying computer tomography technology (Bromley pers. comm.).

A related 'experimental' study carried out by the same authors (Bromley & Asgaard 2004), covers a total exposure time of ~2,100 years (!). In this case the 'test blocks' were deployed by accident, when a marble-statue-laden Greek ship sunk in shallow waters (~25 m) off the Island of Antikythira (Greece) more than two millennia BP. Part of the load was salvaged during the past century by archaeologists and can now – unfortunately only visually – be investigated at the national Museum of Archaeology in Athens. While some of the marble statues lost 10 to 15 cm of marble, others were still pristine where they were partly or completely covered by sediment.

Inter- and supratidal geomorphological experiments

In the intertidal and supratidal zones, there is a certain degree of overlap with geomorphological studies and experiments, some of which also use rock tablets in order to quantify weathering (including bioerosion). A recent multi-methodological study in this context has been undertaken by Naylor & Viles (2002) in the intertidal zone of the Island of Crete (Greece). A review on weathering rock block trials was given by Moses (2000) and a review on littoral biokarst formation in general was provided by Schneider & Torunski (1983).

Summary

When summarising the experimental sites and approaches, we clearly see, that previous bioerosion experiments were (1) with few exceptions limited to tropical and subtropical seas, (2) mostly conducted in reef settings, (3) as a consequence chiefly limited to the uppermost water column and (4) often only taking macroborers and grazers into account. The complete lack of quantitative studies from cold-temperate to polar settings was an important motivation for launching the Kosterfjord experiment.

1.4 Previous experimental carbonate accretion experiments

Many of the bioerosion experiments outlined in the previous section also took carbonate accretion into account in order to balance both processes in relation to each other (for example Kiene 1988; Hassan 1998; Pari et al. 1998, 2002). Numerous other studies exclusively focused on the settlement and recruitment patterns of various hardground communities and thereby employed various types of artificial hard substrates ranging in scale from planted shells to oil platforms (Crisp 1974; Schuhmacher 1977; Cha & Bhaud 2000; Bram et al. 2005). A multitude of environmental controls were investigated such as sedimentation rates and light availability (Maughan 2001), nutrient supply (Tomascik 1991), substrate type (Harriott & Fisk 1987), substrate topographic heterogeneity (Pech et al. 2002) or substrate roughness (Gunkel 1997).

Foraminiferal settlement experiments

While most experiments were directed towards macroinvertebrates such as scleractinians, balanids, serpulimorphs and bryozoans, the present contribution lays its focus on benthic foraminiferans which are an ubiquitous compound in virtually all marine settings and have been widely approved as valuable (palaeo)environmental indicators (reviewed by Murray 1991). Previous experimental studies specifically dealing with the colonisation by benthic foraminiferans are scarce and were chiefly limited to soft-bottom substrates. Buzas et al. (1989) and Buzas (1993) for instance investigated shallow-water artificial sandy substrates in the Indian River and mud-filled boxes at 125 m off Florida. Kitazato (1995) carried out colonisation experiments at a deep-sea site off Japan, and Schafer et al. (1996) studied the foraminiferal temperature sensitivity in heated versus non-heated sand-filled trays in the Bedford Basin (E Canada). In the context of the present study, the experiments by Wefer & Richter (1976) and Wefer et al. (1987) in the western Baltic Sea employing artificial gravel, sand and clay substrates at various water depths and the study by Alve (1999) and Alve & Olsgard (1999) applying colonisation sediment boxes in the Oslofjord are of special interest for evaluating differences in soft versus hard substrate colonisation patterns and rates. Artificial hard substrates were employed by Fujita (2004) in coral reef sites of Japan, by Van Dover et al. (1988) and Mullineaux et al. (1998) near a hydrothermal vent in the abyssal East Pacific, and by Bertram & Cowen (1999) at the deep-sea Cross Seamount site (south of Hawaii, Pacific Ocean). The foraminiferal colonisation on artificial and natural seagrass leaves was studied in detail by Ribes et al. (2000) in shallow waters of Medes Island (NW Mediterranean Sea). Colonisation of new habitats by benthic foraminiferans in general was recently reviewed by Alve (1999).

Benthic foraminiferans in the Skagerrak

Recent benthic foraminiferans have been intensively studied in the Skagerrak (e.g., Höglund 1947; Van Weering & Qvale 1983; Seidenkrantz 1993; Alve & Murray 1995, 1999). The analysis of surface sediments comprises the distribution of distinct assemblage groups related to specific water depths and water masses. Crucial factors that influence their composition are salinity, temperature, water depth, sediment texture plus composition, food availability, and oxygenation (Bergsten et al. 1996). Their distribution has also been used to infer taphonomic processes like transport and dissolution of tests, whereas high benthic fertility is linked with high abundance of particulate organic matter (Alve & Murray 1997). The temporal variation of benthic foraminiferans is the subject of detailed spatial investigations (Moodley et al. 1993; Alve & Murray 1995; Alve 1996). In these studies, the long-term change of benthic foraminiferal assemblages is documented in downcore variations and the comparison of surface sediment samples collected some 50 years earlier (Höglund 1947).

All studies mentioned above concentrated on benthic foraminiferal assemblages of sediment surface samples of the Skagerrak basin. The investigation of shallow-water benthic foraminiferans in coastal settings of Scandinavian fjords, however, is limited, but has been increasingly tackled during the past decades (for example Alve & Nagy 1986; Austin & Sejrup 1994; Alve 1995; Alve & Murray 1999; Gustafsson & Nordberg 1999; Klitgaard-Kristensen & Buhl-Mortensen 1999; Murray & Alve 1999). In many fjord settings the faunal composition of benthic foraminiferans is affected by environmental influences like the pronounced thermohaline stratification patterns, the seasonal varying freshwater supply, seasonal sea ice cover, the dissolved oxygen concentration or seasonal phytoplankton blooms. For example Gustafsson & Nordberg (1999) studied the response of benthic foraminiferans to hydrography, hypoxic conditions and primary production in the Koljö fjord on the Swedish west coast located south of the Kosterfjord. However, non of these investigations deals with the temporal or spatial distribution or settlement of epibenthic foraminiferal hardground communities.

1.5 Agents of high-latitude bioerosion

Tables 2 to 4 provide a compilation of grazing, macroboring and microboring agents reported from high-latitude settings. It needs to be stressed, that the listed related ichnogenera are not necessarily exclusively produced by the quoted biotaxa, and, more than one ichnotaxon may be produced by one and the same biotaxon (see Sect. 1.7).

Table 2 Grazing and crushing bioerosive agents reported from high-latitude, cold-temperate and polar marine settings with the related ichnotaxa and mode of bioerosion

Agent	Important genera	Related ichnogenera	Mode of erosion	Selected references
Gastropods	*Acmaea* *Littorina*	*Radulichnus* *Radulichnus*	Mechanical + chemical rasping	Ankel 1936, 1937; Boekschoten 1966; Abbott 1974; Voigt 1977; Farrow & Clokie 1979; Jüch & Boekschoten 1980; Bromley & Hanken 1981; Akpan 1984; Akpan & Farrow 1985; Farrow & Fyfe 1988; Moen & Svenson 2004
Chitons	*Lepidopleurus* *Lepidochitona*	*Radulichnus* *Radulichnus*	Mechanical + chemical rasping	Boekschoten 1966; Voigt 1977; Farrow & Clokie 1979; Jüch & Boekschoten 1980; Bromley & Hanken 1981; Akpan 1984; Farrow et al. 1984; Akpan & Farrow 1985; Farrow & Fyfe 1988; Moen & Svenson 2004

Table 2 continued

Agent	Important genera	Related ichnogenera	Mode of erosion	Selected references
Echinoids	*Echinus* *Psammechinus* *Paracentrotus* *Strongylocentrotus*	*Gnathichnus* *Gnathichnus* *Gnathichnus* *Gnathichnus*	Mechanical rasping	Milligan 1916; Jehu 1918; Otter 1932; Krumbein & Van der Pers 1974; Bromley 1975; Bromley & Hanken 1981; Akpan 1984; Farrow et al. 1984; Trudgill et al. 1987; Moen & Svenson 2004
Asteroids	*Asterias* *Astropecten*	- -	Mechanical crushing or ingestion	Hyman 1955; Carter 1968; Moen & Svenson 2004
Decapods	*Carcinus* *Cancer* *Homarus* *Hyas* *Pagurus*	- - - - -	Mechanical crushing	Orton 1926; Carter 1968; Noble et al. 1976; Moen & Svenson 2004
Birds	*Haematopus* *Somateria* *Tadorna* *Larus*	- - - -	Mechanical crushing + ingestion	Dewar 1908; Drinnan 1957; Hancock & Urquhart 1965; Wilson 1967; Carter 1968; Farrow 1974; Trewin & Welsh 1976; Cadee 1994
Fish	*Asemichtys* *Pleuronectes*	- -	Mechanical crushing + ingestion	Blegvad 1925, 1930; Dawes 1931; Carter 1968; Norton 1988; Moen & Svenson 2004
Mammals	*Odebenus* *Phoca*	- -	Mechanical crushing	Scheffer 1958; Harrison & King 1965; Carefoot 1977

Table 3 Macroboring bioerosive agents (trace diameters >100 μm) of high-latitude, cold-temperate and polar marine settings with the related ichnotaxa and mode of bioerosion

Agent	Important genera	Related ichnogenera	Mode of erosion	Selected references
Bivalves	*Gastrochaena* *Petricola* *Anomia*	*Gastrochaenolithes* *Gastrochaenolithes* *Centrichnus*	Mechanical + chemical boring + attachment	Jehu 1918; Purchon 1955; Ansell 1970; Abbott 1974; Kelly & Bromley 1984; Trudgill 1987; Trudgill & Crabtree 1987; Bromley & Martinell 1991; Savazzi 1999; Moen & Svenson 2004
Echinoids	*Paracentrotus* *Strongylocentrotus*	*Circolites* *Circolites*	Mechanical boring	Warme 1975
Cephalopods	*Octopus*	*Oichnus*	Mechanical boring predation	Carter 1968; Bromley 1993
Gastropods	*Natica* *Urosalpinx* *Nucella* *Capulus* *Pedicularia* *Patella* *Cellana* *Spiroglyphus*	*Oichnus* *Oichnus* *Oichnus* *Lacrimichnus* *Lacrimichnus* *Lacrimichnus* *Lacrimichnus* *Renichnus*	Mechanical + chemical boring predation or attachment	Jehu 1918; Carriker et al. 1963; Carriker & Yochelson 1968; Carter 1968; Carriker & Van Zandt 1972; Abbott 1974; Bromley 1981; Kelletat 1986, 1988; Aitken & Risk 1988; Young & Nelson 1988; Bouchet & Warén 1993; Moen & Svenson 2004; Morton 2004; Bromley & Heinberg 2006

Table 3 continued

Agent	Important genera	Related ichnogenera	Mode of erosion	Selected references
Brachiopods	*Macandrevia* *Terebratulina* *Hemithyris* *Dallina*	*Podichnus* *Podichnus* *Podichnus* *Podichnus*	Chemical attachment	Ekman 1896; Bromley & Surlyk 1973; Moen & Svenson 2004
Sponges	*Cliona* *Aka*	*Entobia* *Entobia*	Chemical + mechanical boring	Boekschoten 1966; Bromley 1970; Bromley & Hanken 1981; Henderson & Styan 1982; Hoeksema 1983; Farrow et al. 1984; Akpan & Farrow 1985; Young & Nelson 1985, 1988; Trudgill 1987; Farrow & Fyfe 1988; Boerboom et al. 1998; Freiwald & Wilson 1998; Beuck & Freiwald 2005
Phoronids	*Phoronis*	*Talpina*	Chemical boring	Lönöy 1954; Silén 1956; Voigt 1975; Bromley & Hanken 1981; Farrow et al. 1984; Akpan & Farrow 1985; Moen & Svenson 2004
Polychaetes	*Polydora* *Boccardia* *Boccardiella* *Dodecaceria*	*Caulostrepsis* *Caulostrepsis* *Caulostrepsis* *Caulostrepsis*	Mechanical + chemical boring	Boekschoten 1966; Blake 1969; Blake & Evans 1973; Van der Pers 1978; Hoeksema 1983; Farrow et al. 1984; Young & Nelson 1985, 1988; Aitken & Risk 1988; Sato-Okoshi & Okoshi 1997; Moen & Svenson 2004
Echiurids	*Thalassema*	-	?Mechan. boring	Farran 1851; Jehu 1918
Sipunculans	*Sipunculus* *Dendrostomum*	*Palaeosabella* *Palaeosabella*	Mechanical + chemical boring	Jehu 1918; Morton & Miller 1968
Flatworms	*Parvatrema*	*Oichnus*	Chemical boring + embedment	Ruiz & Lindberg 1989
Forami-niferans	*Hyrrokkin* *Cibicides* -	- - '*Semidendrina*-form'	Chemical attachment or boring predation	Cerchi & Schroeder 1991; Cedhagen 1994; Freiwald & Schönfeld 1996; Vénec-Peyré 1996; Bromley et al. (in press)
Cirripeds	*Trypetesa* *Cryptophialus* *Ulophysema* *Balanus* *Verruca*	*Rogerella* *Rogerella* *Oichnus* *Anellusichnus* *Centrichnus*	Mechanical + chemical boring or attachment	Berndt 1903; Brattström 1936, 1937; Tomlinson 1953, 1969, 1987; Boekschoten 1966; Seilacher 1969; Newman & Ross 1971; Lambers & Boekschoten 1986; Bromley 1970; Radwański 1977; Moen & Svenson 2004
Macroalgae	*Macrocystis* *Nerocystis* *Laminaria* *Fucus* *Egregia*	- - - - -	Chemical attachment + mechanical detachment	Emery & Tschudy 1941; Emery 1963; Barnes & Topinka 1969; Sergeant 1975; Kelletat 1986; Bennett et al. 1996; Bromley & Heinberg 2006

Table 4 Microboring bioerosive agents (trace diameters <100 μm) of high-latitude, cold-temperate and polar marine settings with the related ichnotaxa and mode of bioerosion

Agent	Important genera	Related ichnogenera	Mode of erosion	Selected references
Chlorophytes	*Ostreobium* *Eugomonita* *Phaeophila* *Gomontia* *Epicladia*	*Ichnoreticulina* *Rhopalia* *Rhopalia* *Cavernula* *Eurygonum*	Chemical boring	Bornet & Flahault 1888, 1889; Kylin 1935; Pia 1937; Kornmann 1959, 1960, 1961; Boekschoten 1966; Wilkinson & Burrows 1970, 1972; Nielsen 1972, 1987; Wilkinson 1974, 1975; Golubic et al. 1975; Kobluk & Kahle 1978; Kornmann & Sahling 1980; Bromley & Hanken 1981; Henderson & Styan 1982; Akpan & Farrow 1984a and b, 1985; Farrow et al. 1984; Akpan 1986; Nielsen 1987; Trudgill 1987; Young & Nelson 1988; Nielsen et al. 1995; Glaub et al. 2002
Rhodophytes	*Porphyra* *Bangia*	'*Palaeoconchocelis*'[A] '*Palaeoconchocelis*'[A]	Chemical boring	Pia 1937; Nielsen 1972, 1987; Conway & Cole 1977; Kobluk & Kahle 1978; Clokie et al. 1979, 1981; Clokie & Boney 1980; Henderson & Styan 1982; Akpan & Farrow 1984a and b; Nielsen et al. 1995; Glaub et al. 2002
Cyano-bacteria	*Hyella* *Solentia* *Mastigocoleus* *Plectonema*	*Fascichnus* *Fascichnus* *Eurygonum* *Scolecia*	Chemical boring	Bornet & Flahault 1888, 1889; Pia 1937; Nielsen 1972, 1987; Golubic et al. 1975; Henderson & Styan 1982; Young & Nelson 1988; Glaub et al. 2002
Bacteria	-	*Scolecia*	Chemical boring	Henderson & Styan 1982; Young & Nelson 1988; Freiwald 1995, 1998
Fungi	*Ostracoblabe* *Dodgella* *Phytophthora* *Schizochytrium* *Conchyliastrum*	*Orthogonum* *Saccomorpha* *Saccomorpha* '*Flagrichnus*-form' *Polyactina*	Chemical boring	Pia 1937; Zebrowski 1937; Höhnk 1969; Kohlmeyer 1969; Cavaliere & Alberte 1970; Henderson & Styan 1982; Akpan & Farrow 1985; Young & Nelson 1988; Boerboom et al. 1998; Freiwald & Wilson 1998; Glaub et al. 2002; Beuck & Freiwald 2005
Bryozoans	*Penetrantia* *Immergentia* '*Spathipora*'[B] '*Terebripora*'[B] *Electra*	*Iramena* - *Pennatichnus* *Pinaceocladichnus* *Leptichnus*	Chemical + mechanical boring or attachment	Zebrowski 1937; Silén 1946, 1947, 1956; Boekschoten 1966; Soule & Soule 1969; Pohowsky 1974, 1978; Taylor et al. 1999; Beuck & Freiwald 2005
Flatworms	*Pseudostylachus*	*Oichnus*	Chemical boring	Woelke 1957; Yonge 1963, 1964
Diatoms?	*Diploneis*? *Cocconeis*?	- -	Chemical	Farrow et al. 1978

[A] No valid ichnotaxon
[B] Traces described as biotaxa

1.6 Agents of microbioerosion

In the following account, a brief overview on the ecology, systematics, distribution and fossil record of the main agents of microbioerosion is given. For corresponding information on macroborers and grazers, the reader is referred to the review papers by Warme (1975) or Bromley (1970, 2004) and the references listed in the tables in the previous chapter (Tables 2-4).

Several microhabitats within the lithobiontic ecological niche are to be distinguished, only one of which addresses actively penetrating microboring organisms, the 'euendoliths'. Organisms which inhabit pre-existing fissures are termed 'chasmoendoliths' and when dwelling within pre-existing cavities 'cryptoendoliths' (Golubic et al. 1981). Organisms settling on the external surface of the rock are termed 'epiliths'. However, epilithic organisms may also have euendolithic parts (e.g., endolithic rhizoids of the epilithic algae *Acetabularia*) or may produce euendolithic stages in the course of their life cycle (e.g., the *Codiolum*-stages of some rhodophytes). Hence these boundaries are not always clearly drawn.

The fossil record of microborings is extensive and dates back as far as the Early Proterozoic (Glaub & Vogel 2004; Bromley 2004). Even though only few fossil examples are known where organic remains of the endolithic organisms have been preserved within their boring systems, most microborings perfectly match the body outline of the producing endoliths and can thus be assigned to their living tracemaking counterparts with a fair degree of confidence. Thus, neontological studies are crucial for providing the linkage between biotaxa and the related ichnotaxa as a basis for the interpretation of fossil traces and palaeoecological applications (see Sects. 1.7 and 1.8).

Eubacteria and Archaea

As for the bacteria, only the cyanobacteria have received major attention (see below) and are among the most important agents of microbioerosion in all photic environments. Other bacteria, both of the domains (Eu)bacteria and Archaea are ubiquitous in marine environments as well – and most likely also contribute to bioerosion of calcareous substrates as hypothesised already by Friedman et al. (1971). However, they are difficult to address not only owing to their minute size. From the ichnological perspective, certain very small traces in fossil and Recent shells have been tentatively assigned to the work of bacteria, the most common one of which being the meandering traces *Scolecia serrata* and *Scolecia maeandria* (e.g., Budd & Perkins 1980; Young & Nelson 1988; Radtke 1991). The fossil record of the latter reaches back to Triassic times (Glaub & Vogel 2004). Also, minute globular or filamentous trace fossils in a sub μ-scale were repeatedly recorded in epoxy resin casts but are difficult to address as boring traces with certainty (e.g., 'Coccoidal Form' in Beuck & Freiwald 2005 or 'Type B1' in Henderson & Styan 1982).

The general importance of bacteria inducing carbonate dissolution by the degradation of skeletal binding organic matter (maceration) was stressed by Freiwald (1995, 1998). In this case, however, the carbonate dissolution is a metabolic by-product rather than an actively boring habit of the bacteria, but nevertheless represents a bioerosive process. The study of bioeroding bacteria is very premature to date and major advances can be expected from future studies combining high resolution SEM with molecular genetic methods such as fluorescence in situ hybridisation (FISH) and others.

Cyanobacteria

The cyanobacteria (formerly also known as chlorophyta, blue-green-algae or blue-algae) support a long living phylogenetic and systematic uncertainty since they exhibit features both of bacteria (prokaryotic cell structure) as well as eukaryotic algae (differentiated heterocysts, photosynthesis). Based on nucleic acid analyses, they are currently placed in the domain of bacteria. The cyanobacteria are primarily obligate photoautotrophs but facultative photoheterotrophy and even chemoheterotrophy has been reported for certain (non-endolithic) species (e.g., Rippka 1972; May & Perkins 1979). For endolithic species a facultative heterotrophy is not confirmed but is discussed for *Plectonema terebrans* which thrives well under dysphotic conditions (Glaub 1994). The reproduction is limited to varying vegetative strategies.

The cyanobacteria bear several dozen endolithic species, the most prominent of which being representatives of the genera *Hyella*, *Solentia*, *Mastigocoleus* and *Plectonema* (e.g., Gektidis 1997b). Important respective ichnospecies are *Eurygonum nodosum*, *Scolecia filosa* and various species of *Fascichnus* and *Planobola*. Endolithic cyanobacteria dominate supra- to shallow sublittoral boring communities but are also found down to dysphotic depths (deepest record of *Plectonema terebrans* in 370 m water depth; Lukas 1978). As boring mode, both chemical dissolution via carbonic acid formed from metabolic CO_2 (Schneider 1976) as well as the support by endosymbiontic bacteria are discussed (Lukas & Golubic 1981). Most recently, Garcia-Pichel (2006) proposed a 'calcium pump' - the active extrusion of calcium ions through an active cellular uptake and transport process - as most appropriate model for the mode of penetration.

Fossil endolithic cyanobacteria and/or their borings mark the earliest record of the endolithic life habit by dating back as far as the Early Proterozoic with the oldest records of *Euhyella campbellii* reported from 1.7 billion year old stromatolites discovered in China (Zhang & Golubic 1987). Some species of *Euhyella* that were already present during Proterozoic times still occur in the Recent (*E. dichotoma* and *E. rectoclada*) and thereby give good examples for the longevity of many microendoliths (Glaub & Vogel 2004).

Fungi

Endolithic marine fungi are ubiquitous chemoheterotrophic bioerosion agents, which penetrate calcareous shells in order to exploit mineralised organic matter (organic lamellae, conchiolin, chitin) or other endolithic organisms for nutrition (Jones & Pemberton 1987; Golubic et al. 2005). Consequently, they are independent of light and can be found down to abyssal water depths both as primary as well as secondary colonisers. However, fungi also engage in parasitic, saprophytic and symbiotic relationships and therefore may have an indirect affinity to light environments (Golubic et al. 2005). The most prominent symbiotic interactions of fungi with cyanobacteria and/or algae characterise another bioerosion agent, the lichens, which in marine environments are only of importance in the eulittoral and especially supralittoral, where a specific zonation pattern can be recognised (Schneider 1976).

An endolithic habit is known for 16 genera (Zeff & Perkins 1979: table 1) among the higher fungi (Eumycetes) and lower fungi (Phycomycetes). Reproduction takes place by vegetative as well as sexual means. The systematic classification of endolithic fungi (and of fungi in general) is still premature owing to major culturing difficulties, opening a wide field for future studies applying DNA-sequencing for phylogenetic assessments.

Fungi employ organic acids such as citric acid for penetration (Jones & Pemberton 1987), but studies in this respect are very limited as yet. Golubic et al. (2005) suggest the presence of specific enzymes for digestion purposes of the organic matter, allowing also for an unhindered penetration of shells rich in organic lamellae.

Marine endolithic fungi make their appearance in the fossil record in the earliest Palaeozoic where Chytridiomycetes were reported from Cambrian shells (Müller & Löffler 1992). The first appearance of the important ichnotaxa *Orthogonum fusiferum* and *Saccomorpha clava* is recorded from the Ordovician and for *Polyactina areneola* from the Silurian (Glaub & Vogel 2004).

Chlorophytes

Green algae (Chlorophyta) are eukaryotes which reproduce either by vegetative or sexual means and exhibits chloroplasts with chlorophyll a and/or b. Only from a dozen genera of the orders Ulotrichales, Chaetophorales and Caulerpales, endolithic forms are known that live entirely endolithic or develop endolithic *Codiolum*-stages during their life cycle. Chlorophytes are photoautotrophs and consequently dependant on light irradiance for photosynthesis. For *Ostreobium quekettii*, which is known to cope with very low light availabilities, a facultative chemoheterotrophy was suggested by Schroeder (1972) and disputed by Glaub (1994). Endolithic chlorophytes are assumed to bore by chemical means by utilising the CO_2 produced during night-time respiration as carbonic acid for the dissolution of the calcareous substrate (e.g., Schneider 1976). Some endolithic algae are known to live in a symbiosis relationship with scleractinian corals as for instance reported for *Ostreobium quekettii* by Schlichter et al. (1997).

Green algae appear during the Cambrian and the oldest endolithic form was reported from the Ordovician represented by the ichnotaxon *Ichnoreticulina elegans*. The important taxa *Cavernula pediculata* as well as *Rhopalia catenata* are known since the Triassic (Glaub & Vogel 2004).

Rhodophytes

Among the eukaryote red algae (Rhodophyta) only few endolithic representatives are known. The Bangiaceae with the most prominent genera *Porphyra* and *Bangia* for instance are known to follow a complex heteromorphic life cycle and develop endolithic *Conchocelis*-stages (Campbell 1980). As the only respective ichnotaxon, '*Palaeoconchocelis starmachii*' was introduced to the literature (Campbell et al. 1979) but confusingly as a mixture of body- and trace fossil (and thus written in quotation marks here). Only premature studies have been directed towards endolithic rhodophytes and their traces yet, implying the potential for further discoveries in this respect.

The fossil record of rhodophytes reaches as far back as the Proterozoic, documented by body fossils of *Bangia* discovered in 1.2 billion years old strata in Arctic Canada (Butterfield et al. 1990). The oldest record of the *Porphyra*-related boring trace '*Palaeoconchocelis starmachii*' stems from Silurian strata (Campbell 1980; Glaub & Bundschuh 1997) and a potential further rhodophyte ichnotaxon (*Orthogonum tripartitum*) was recorded in Ordovician ostracode valves (Olempska 1986; Schmidt 1992).

Bryozoans

Among the diverse group of the bryozoans, more than a dozen genera, most of which belonging to the order Ctenostomata, follow an endolithic mode of life. Ctenostome bryozoans do not develop a calcareous exoskeleton, explaining their ecological advantage of boring and dwelling in calcareous skeletons as protective housings. Pohowsky (1978) provided the most extensive investigations of these bioerosion agents to date and describes or notes 48 endolithic forms. There is an unresolved disputation on the formal status as biotaxa versus ichnotaxa of the most common species and specifically of *Spathipora, Penetrantia* and *Terebripora* (e.g., Boekschoten 1970; Pohowsky 1974, 1978; Mayoral 1988). They were originally established as biotaxa but were solely based on the morphology of their borings and should thus be regarded as ichnotaxa. Subsequently, Boekschoten (1970) and Mayoral (1988) offered the respective ichnotaxa *Pennatichnus*, *Iramena* and *Pinaceocladichnus*, which are, however, not widely appreciated as yet. All of these forms have the development of individual zooids, interconnected by a tubular system of stolons in common. From the initial zooid (ancestrula), stolons develop interconnecting the subsequent zooids (autozooids and/or heterozooids) which may be connected to the stolons by short tunnels (pedunculus). Primarily chemical penetration aided by phosphoric acid (Silén 1947) as well as a mechanical aided boring (Borg 1940) have been suggested as likely boring mechanisms.

In contrast to the ctenostome bryozoans, a number of cheilostome genera etch shallow attachment scars (ichnogenus *Leptichnus*) when encrusting calcareous substrates such as the genus *Electra*. In this case, boring is most likely chemical and may result in better adherence to the substrate, giving protection from physical abrasion and bioerosion (Taylor et al. 1999).

The stratigraphical range of the various families and genera was given by Pohowsky (1978) with *Ropalonaria* being the oldest representative with a Lower Ordovician emergence. The common Spatiporiidae and Penetrantiidae appear during the Mesozoic and the Terebriboridae and Immergentiidae during the Eocene and Miocene, respectively.

1.7 Ichnotaxonomy and biotaxonomy

When degrading calcareous substrates, bioerosion agents inevitably modify the substrate and leave more or less prominent traces of their action. These features have – in contrast to soft bottom traces – a high potential of becoming preserved in the fossil record and can be regarded as 'incipient' trace fossils.

The bioeroding organisms are taxonomically classified as biotaxa either within the International Code of Botanical Nomenclature (ICBN 2000) or the International Code of Zoological Nomenclature (ICZN 1999). Their traces are classified as ichnotaxa and are covered by the ICZN as well since the specific amendments of the ichnological provisions to the code in the year 1985 (Rindsberg 1990). Consequently, a dual nomenclature as proposed for instance by Bromley & Fürsich (1980) and Bromley (1981) is applied herein, with the zoological or botanical taxa for classifying the bioeroding organisms, and the ichnotaxa as a morphological classification of the traces they leave. Thereby the basic principles of ichnology are acknowledged (after Bromley & Fürsich 1980):

- Trace fossils are structures produced in sediments and hard substrates by the activity of organisms.
- The nomenclature of trace fossils is based solely upon the morphological characteristics of the structure.
- A particular structure may be produced by the work of two or several different organisms living together, or in succession, within the structure.
- The same individual or species of organism may produce different structures corresponding to different behavioural patterns, or to identical behaviour but in different substrates, respectively.
- Identical structures may be produced by the activity of systematically different trace-making organisms, where behaviour is similar.

For further information on ichnotaxonomic concepts and principles the reader is referred to the comprehensive reviews by Magwood (1992), Pickerill (1994) and Bromley (1996).

To date, the ICZN permits the use of ichnotaxa only for fossil borings. This restriction concerning neoichnology is a matter of ongoing debate (e.g., Bromley & Fürsich 1980; Rindsberg 1990; Magwood 1992; Tavanier et al. 1992; Pickerill 1994; Bromley 1996; Nielsen et al. 2003; Radtke & Golubic 2005). The limitation is not expedient in the case of traces in hard substrates for a number of reasons such as the lack of an applicable definition regarding the fossilisation threshold, or the fact that a Recent boring morphologically resembles its fossil counterpart in all aspects. In the present volume, ichnotaxa are employed as it has been suggested for instance by Bromley & Fürsich (1980) and Rindsberg (1990) and has practically been done by numerous workers before, (e.g., Radtke 1993; Vogel et al. 2000;

Perry & MacDonald 2002; Glaub 2004; Radtke & Golubic 2005). However, where new ichnotaxa were introduced (e.g., *Eurygonum pennaforme* Wisshak, Gektidis, Freiwald & Lundälv 2005), fossil holotype material was provided confirming the rules of the ICZN. Where no fossil equivalents were obtainable, or observations are too limited as yet for justifying the establishment of a new ichnotaxon, open nomenclature was applied (e.g., 'Sponge-form I').

1.8 Bioerosion as a palaeoenvironmental tool

Keeping the obligatory reservations concerning the actualistic approach in mind (e.g., the possibility of evolutionary changes in ecological demands of certain biota), observations regarding the physiological and ecological demands of microendoliths can help us when judging not only relative palaeobathymetry but also palaeosalinities and palaeotemperatures. Both temperature as well as salinity may be buffered to some extent in the microenvironment of the endolithic cavities (Wilkinson 1974). This effect makes endoliths somewhat resistant concerning short-term environmental fluctuations and is reflected to some extend by the evolutionary longevity of endolithic organisms in general (Vogel & Glaub 2004).

Palaeobathymetry

For geologists the reconstruction of ancient water depths (palaeobathymetry) for a depositional setting or palaeocommunity is fundamental to reconstruct palaeoenvironments – but is probably the hardest parameter to measure. The investigation of microendolithic trace communities has been shown to be a strong tool for judging light availability and hence relative bathymetry in both modern and ancient environmental settings. During the past decade, a set of bathymetric index ichnocoenoses was established (Table 5), characterising the shallow-euphotic II (eulittoral) + III (sublittoral), deep-euphotic and aphotic zones (Glaub 1994, 1999; Vogel et al. 1995, 1999; Glaub et al. 2002; Vogel & Marincovich 2004). Index ichnocoenoses for the shallow-euphotic I (supralittoral) and the dysphotic zones are yet to be defined. Extensive calibration studies in various fossil and in modern tropical and subordinately high-latitudinal settings showed a remarkable consistency in this ichnocoenoses zonation pattern and the scheme has been approved as valuable palaeobathymetric guidline for numerous Phanerozoic case studies (Vogel et al. 1995).

Table 5 The photic zonation as reflected by specific index ichnocoenoses and general microborer distribution characteristics (slightly modified after Glaub 1994, 1999 and Vogel & Marincovich 2004).

Photic zonation		**Index ichnocoenoses**	**General characteristics**
Euphotic zone (>1% surface illumination)	shallow I (supralittoral)	not yet defined	Dominance of cyanobacteria with sheath pigmentation
	shallow II (eulittoral)	*Fascichnus acinosus* / *Fascichnus dactylus* ichnocoenosis	Dominance of cyanobacteria; vertical orientation of borings
	shallow III (sublittoral)	*Fascichnus dactylus* / '*Palaeoconchocelis starmachii*' ichnocoenosis	Cyanobacteria abundant and eukaryotes; change from vertical to horizontal orientation
	deep	'*Palaeoconchocelis starmachii*' / *Ichnoreticulina elegans* ichnocoenosis	Dominance of eukaryotes; mainly rhodophytes and chlorophytes; horizontal orientation; heterotrophs increasing; maximum diversity
Dysphotic zone (0.01-1% surface illumination)		not yet defined	Dominance of heterotrophs; additionally *Ichnoreticulina elegans* and/or *Scolecia filosa*
Aphotic zone (<0.01% surface illumination)		*Saccomorpha clava* / *Orthogonum lineare* ichnocoenosis	Only heterotrophs

Palaeotemperature

Even though the potential suitability of microendolithic traces for palaeotemperature and/or linked palaeolatitude judgments was long recognised (e.g., Golubic et al. 1975), they have been rarely applied as such, implying the need for future work on this promising issue.

In modern marine settings, many microendoliths exhibit a cosmopolitan distribution, but a number of taxa occur within certain temperature limits and are consequently restricted for instance to the Tropics and Subtropics. Most endolithic algae for example exhibit similar upper limits in their temperature resistance (~35°C) but show significant differences concerning their lower temperature limit (Lüning 1985). Consequently, the absence of certain endolithic algae or their corresponding traces can be utilised as temperature indicators for different non-tropical climates (e.g., *Acetabularia* sp. / *Fascichnus grandis*; Glaub et al. 2002). The same applies to certain cyanobacteria (e.g., *Hyella racemus* / *Fascichnus rogus*, *Mastigocoleus testarum* / *Eurygonum nodosum*, *Cyanosaccus piriformis* / *Planobola* isp.; Lukas & Golubic 1981; Bundschuh & Balog 2000; Glaub et al. 2002) and fungi (e.g., *Saccomorpha sphaerula* / *Lithophytium* sp.; Glaub et al. 2002).

However, palaeoenvironmental interpretations founded on the absence of certain taxa is based on negative evidence and thus of limited reliability, implying the importance to identify taxa that indicate cold-water conditions by their presence. Potentially useful candidates in this respect are the ichnospecies *Saccomorpha terminalis*, produced by the marine fungus *Phytophthora* and the trace '*Flagrichnus*-form 2', the boring of an unknown heterotroph (Wisshak & Porter in press), the latter one of which having been recognised and described in the course of the Kosterfjord experiment.

Palaeosalinity

Like for palaeotemperature, microendoliths bear the potential as palaeo-salinity indicators but have to date not been widely appreciated as such. Only few studies tackled salinity tolerances of specific Recent endolithic species as yet, such as the studies on the fungus *Ostracoblabe implexa* (trace: *Orthogonum fusiferum*) by Alderman & Gareth Jones (1971), or the investigations on the salinity dependant morphology variations of *Phaeophila dendroides* and *Eugomontia sacculata* (trace: *Rhopalia catenata*) by Nielsen (1972) and Wilkinson (1974). Further work in this respect - for instance in form of controlled laboratory experiments - is required to broaden the base for future palaeoenvironmental salinity estimations.

1.9 Bioerosion and the global carbon(ate) cycle

The biogenic carbonate accretion and the carbonate breakdown via bioerosion are opposing players and both inevitably linked to the global carbonate as well as carbon cycle, the former being a compound of the latter. They are thereby not limited to the marine realm but are also relevant processes in lacustrine and terrestrial environments. Considering that an annual 3.5 billion tons of marine calcareous sediments are deposited (Schneider et al. 2000) and carbonates make up approximately 17.6% by volume of all sedimentary rocks, thereby representing the principal carbon dioxide source as well as sink by containing an estimated 78 x 10^{12} tons of CO_2 (Mackenzie 1998), a furthered understanding of bioerosion and carbonate accretion is vital not only to the (palaeo)biologist but also to the geologist and (palaeo)climatologist. Carbonates and their relationship to the complex global carbon cycle gained immense attention during the past decades in terms of the role of the greenhouse gas CO_2 in future global climate change (see Schneider et al. 2000, Broecker 2003 and Ridgwell & Zeebe 2005 for recent reviews).

On a global geologic scale, the carbonate mega-cycle consists of an exogenic part taking place in the atmosphere and hydrosphere, comprising the processes of weathering, transport, precipitation, (bio)erosion, sedimentation and early diagenesis, and an endogenic part in the lithosphere, including diagenesis, metamorphosis, orogenesis and epeirogenesis (Fig. 2).

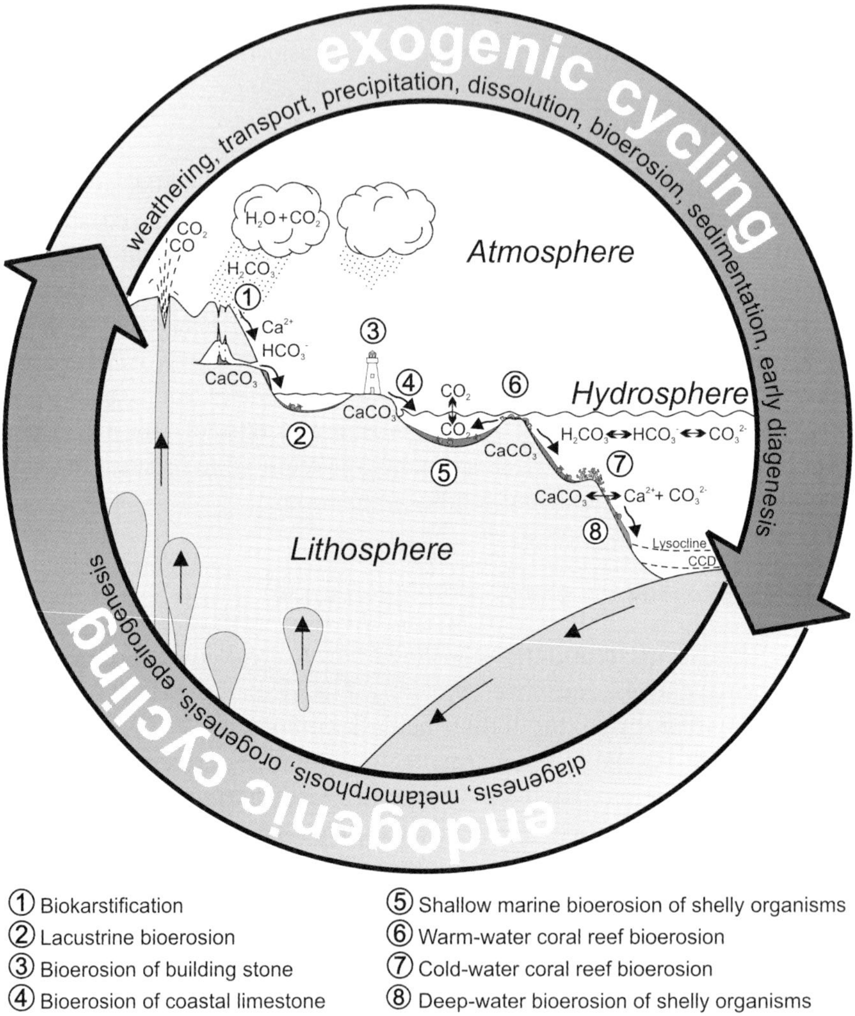

Fig. 2 The inorganic carbon(ate) cycling in the atmosphere, hydrosphere and lithosphere with the main settings for carbonate precipitation and bioerosion, and the major carbon and carbonate species involved (based upon Golubic & Schneider 1979 and Golubic et al. 1979)

Both bioerosion and carbonate accretion are key compounds in the exogenic cycling where carbonate may repeatedly be precipitated and bioeroded before ultimately entering the endogenic part of the cycle by sedimentation and diagenesis (Golubic et al. 1979; Ridgwell & Zeebe 2005). The complex exogenic part in itself consists of a variety of sub-cycles owing to the multitude of environments, processes and carbon(ate) species involved.

Bioerosion contributes to the carbonate breakdown by means of two principal mechanisms: biocorrosion by chemical carbonate dissolution (e.g., most microborers) and bioabrasion by mechanical removal of carbonate (e.g., most grazers). The two mechanisms are often found synergistically interlinked, both in the boring manner of individual bioerosion agents as well as on a biocoenosis level. Grazing for instance is facilitated by prior acidic weakening of the substrate either by the grazer himself or by other bioeroders. Furthermore, the mechanical breakdown increases the surface/volume ratio, promoting enhanced microbioerosion activity and early diagenetic dissolution within the sediments. Ultimately bioerosion either liberates calcium ions Ca^{2+} and bicarbonate ions HCO_3^- to the hydrosphere or contributes to fine-grained carbonate sedimentation. Carbonate accretion on the other hand removes Ca^{2+} as well as HCO_3^- and other carbonate species from the calcite-carbonate equilibrium in the hydrosphere ($H_2O + CO_{2(aq)} \leftrightarrow H_2CO_3 \leftrightarrow H^+ + HCO_3^- \leftrightarrow 2H^+ + CO_3^{2-}$) by incorporation into skeletal carbonate, thereby forming potential new substrate for bioerosion agents. As a consequence for the CO_2 cycling and interlinked climatic relevance, bioerosive dissolution of $CaCO_3$ counter-intuitively drives a pCO_2 decrease in the hydrosphere and by equilibrium processes also in the atmosphere, while carbonate precipitation leads to an increase in ocean pCO_2.

In addition, the bioeroding and carbonate accreting biota influence the carbon and carbonate cycling by means of photosynthesis and/or respiration, again two opposing processes: while photosynthesis decreases the pCO_2 partial pressure and HCO_3^- ion content in the water, leading to a raised pH and CO_3^{2-} concentration and is thus promoting carbonate precipitation, respiration functions in opposite direction, principally promoting carbonate dissolution. Photosynthesis and respiration are furthermore chief components in the equally important organic carbon cycle, driving the biosynthesis and biomineralisation of organic matter $(CH_2O)_n$. Hence, the organic carbon cycling is mutually linked to the inorganic carbon(ate) cycle (e.g., Golubic et al. 1979; Wollast 1994).

When acknowledging the role, bioerosion and carbonate accretion play in the carbon(ate) cycling, the relevance of qualitative and especially of quantitative assessments of bioerosion and carbonate accretion rates as integral compounds for flux modelling and ultimately (palaeo)climatologic reconstructions becomes evident. It is especially interesting in this context, that the most recent marine carbonate budget assessments (Milliman 1993; Wollast 1994; Schneider et al. 2000) yield an imbalance of the marine carbonate budget which could be explained by overestimated fluxes due to carbonate production or underestimated (biocorrosive?) dissolution. Intriguingly, the term 'bioerosion' appears rarely if at all in corresponding models. Only where carbonate budgets for tropical reefs were assessed, bioerosion is a widely considered flux (e.g., Scoffin et al. 1980b; Hassan 1998).

For the cold-temperate to polar realms, the present study is the first to yield quantified data on bioerosion rates and we shall see that there are significant differences in bioerosion rates with latitude (Sect. 6.2). Carbonate accretion budgeting in context of the global carbonate cycle has for instance most recently been undertaken for cold-water coral reefs by Lindberg & Mienert (2005). However, more effort has to be directed towards further budgeting in order to allow a proper comparison of temperate to polar seas to the well studied tropical realm.

1.10 Bioerosion versus physicochemical dissolution

The effects of early sea-floor processes on the taphonomy of skeletal carbonate deposits in high-latitude seas form a complex interplay of various chemical (dissolution, precipitation), physical (abrasion, breakage) and biological (bioerosion, encrustation, bioturbation) processes. The nature and outcome of this interplay leads to very different sediment characteristics and preservation potential if compared to warm-water tropical seas (see Smith & Nelson 2003, for a review).

In general, the dissolution and precipitation of carbonate in the ocean is mainly controlled by the saturation state of sea-water (Ω), which is dependant on the ambient concentrations of Ca^{2+} and CO_3^{2-} and is influenced by temperature and pressure. In this context a long living misinterpretation was introduced to the literature, triggered by the investigations on early seafloor processes by Alexandersson (1974a, 1974b, 1976, 1978, 1979): the myth of generally undersaturated cold waters in the high-latitudes as opposed to the supersaturated warm waters in tropical seas. With respect to the Skagerrak, Alexandersson (1979) consequently considered physicochemical dissolution as much more important than microborers for the destruction of carbonates. This view has to be revised for a number of reasons: The surface water of the world oceans is generally supersaturated with respect to $CaCO_3$ with an average Ω of 4.8 for calcite (Ridgwell & Zeebe 2005). On the neritic continental shelf areas, pure physicochemical dissolution on the seafloor (as opposed to within; see below) is comparatively insignificant (Rao 1996). It is only important in the deep sea towards the CO_3^{2-} saturation horizon (lysocline; $\Omega = 1$) and especially the carbonate compensation depth (CCD), where low temperatures and high pressure of the water column considerably shift the equilibrium in a direction where calcite becomes thermodynamically unstable (Wollast 1994). In the past decade, however, it has been recognised, that a considerable amount of carbonate dissolution occurs in the deep sea well above the lysocline, the reason for which being not yet fully understood (Milliman et al. 1999; Chung et al. 2003).

In contrast to the supersaturated water column, dissolution may well take place below the sediment/water interface – driven by carbonic acid formed as a result of CO_2 released by the respiration of organic matter under aerobic conditions (Wollast 1994). This dissolution is not of pure physicochemical nature but is in fact to a considerable portion driven by the metabolic activity of bacteria (maceration), a process that can be regarded as microbial bioerosion (Freiwald 1995, 1998). Hence, at least part of the dissolution features observed by Alexandersson or subsequent workers such as Cutler & Flessa (1995) on shell material from the Skagerrak may represent bioerosive features rather than signs of pure physicochemical dissolution.

With the present experimental design, substrates placed with some distance above the sediment/water interface were studied. Consequently, it can be assumed with confidence, that physicochemical dissolution did not alter the assessment of bioerosion rates. The insignificance of dissolution is furthermore underlined by the complete lack of dissolution features on a scanning electron microscopic scale. The surfaces of the mollusc substrates are pristine even in high resolution SEMs and all microendolithic traces are perfectly well preserved.

1.11 Objectives of the Kosterfjord experiment

A two-year bioerosion experiment was designed and launched in September 2002 in cooperation with the Tjärnö Marine Biological Laboratory (TMBL) in the cold-temperate setting of the northern Kosterfjord area off SW Sweden. This interdisciplinary study tackled four major and mutually linked goals:

1. Determination of bioerosion and carbonate accretion rates along a bathymetric and hydrographic gradient.

2. Taxonomical identification of the bioerosion agents (chiefly microborers), the traces they leave and the analysis of the ichnocoenoses composition.
3. Taxonomical identification of the calcareous epizoans (chiefly foraminiferans, serpulimorphs and bryozoans) and the analysis of the biocoenoses composition.
4. Comparison of the results to tropical endolithic communities and bioerosion rates and their interpretation in context of the global carbon(ate) cycling.

Accompanying the experimental approach, endolithic colonisation in dead skeletons of the cold-water coral *Lophelia pertusa* from the aphotic zone as well as of the shallow-water gastropod *Littorina littorea* were studied in detail (Sects. 4.3 and 4.4). These investigations on background sample material enable the comparison of the experimental results with advanced taphonomic stages of shell degradation.

The experimental approach bears a number of major advantages when studying bioerosion processes since it (1) eliminates any bathymetric bias caused by transport of shell material, (2) gives information on the orientation of the colonised shells, (3) enables the study of initial stages and the succession of colonisation, and (4) provides a reliable way to quantify bioerosion.

2 Material and methods

2.1 Assessing environmental parameters

Long-term hydrologic record

The general oceanographic watermass stratification and seasonal variability of the study area was evaluated on the basis of an extensive hydrologic dataset (© Swedish Fishery Board) consisting of more than 700 CTD profiles, provided by the Tjärnö Marine Biological Laboratory (TMBL). Measurements were taken near the study site (59°00'4" N / 11°06'8" E) in the northern Kosterfjord between 1967 and 1990.

Short-term hydrologic record

The site specific short-term hydrographic conditions at the Säcken site were obtained through two lander deployments (Aanderaa RCM 9) of plus one month each and were carried out by T. Lundälv (TMBL). The first deployment covered part of the warmer seawater conditions in September to October 2000, whereas part of the colder period was covered by a second deployment in February to April 2001. The autonomous recorder logged high-resolution measurements of temperature, salinity, current direction, current speed and turbidity in one hour intervals.

Light measurements

Light measurements were carried out applying a LICOR Spherical Quantum Sensor (LI-193SA) in combination with a datalogger (LI-1400). The relevant measured spectrum lay within a wavelength range of 400-700 nm, which corresponds to the photosynthetically active radiation (PAR; unit = μmol photons m^{-2} s^{-1}). The recorded data were additionally translated to percentages with respect to surface illumination. Numerous measurements were taken in late September 2002, late March 2003 and late October 2003, always around noon under a wide range of weather conditions from cloudless calm weather to overcast and windy conditions.

2.2 The experimental design

The basic setup

The basic units of the experiment were PVC-frames, 40 x 70 cm in size, with four legs (concrete-filled polypropylene tubes) and a central upright PP-tube with a rope sling for recovery purposes (Fig. 3). Attention was drawn to the selection of the materials used, avoiding corrodible material and glue with anti-fouling agents to prevent distraction of potential bioeroding organisms. In order to withstand enhanced hydrodynamic forces, the 7 m panels were equipped with additional weights (long concrete filled PP-tubes) and the 1 m panels were directly mounted on large concrete blocks (40 x 70 x 30 cm).

For the quantitative analyses, 6 accretion PVC plates and 6 erosion limestone plates (all 10 x 10 x 1 cm) were mounted on the upward facing side of each panel. For the analysis of the microendoliths, different substrates were attached to the frames: small Iceland spar and micrite blocks, as well as cleaned (boiled) fresh bivalve shells and coral fragments.

Fig. 3 A The basic unit of the experiment and the different substrates mounted on it. **B** A set of three experimental panels next to each other on the sea floor at 7 m water depth

Deployment and recovery of panels

The experimental site in the northern Kosterfjord area was chosen with the aid of a navigation- and GIS-software (OLEX) and explored with Phantom XTL – and Phantom S4 ROVs (Deep Ocean Engineering), equipped with headlights, video camera, sonar, navigation system, depth gauge and a laser scaling device. With ROV support, the deployment of the frames was monitored and sediment samples as well as video-footage from each deployment site were gathered. Between September 27 and October 3, 2002, three frames, respectively, were deployed along a bathymetric transect at 1 m, 7 m, 15 m, 30 m, 50 m and 85 m water depth in the northern Kosterfjord area (Fig. 3B). A first subset of panels was recovered with ROV support six months later between March 24 and 27, 2003, and the second set after approximately one year exposure between October 23 and 26, 2003. The third and final recovery was carried out late September 2004, extending the overall exposure period to about two years. An additional set of panels was deployed during the first recovery session at 1 m in a lagoonal setting and was exposed for 6 and 18 months. Only one panel (1 m at the transect after 2 years exposure) was lost due to strong winter storms in combination with ice drift during the winter 2003/2004.

2.3 Preparation and evaluation techniques

In the following account, the various preparation and evaluation techniques employed in the course of the experiment are outlined (Fig. 4).

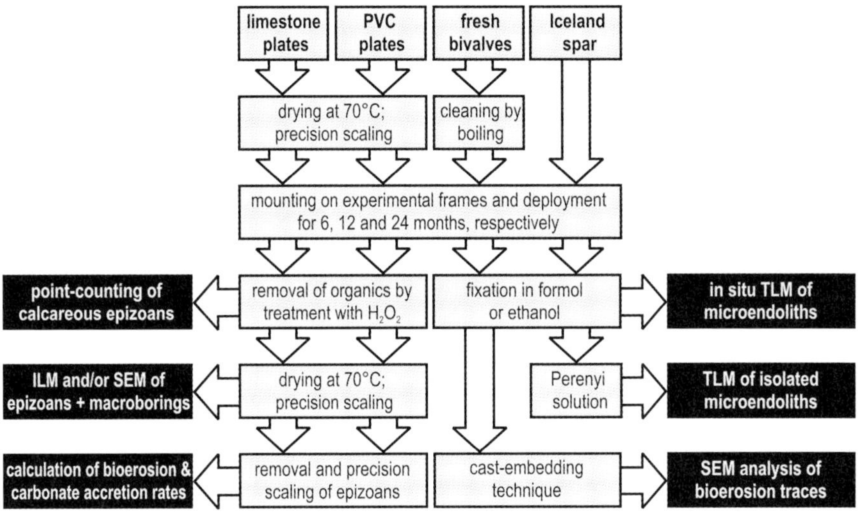

Fig. 4 Flowchart illustrating the main pathway of deployment, preparation and analysis of the different substrates (ILM = incipient light microscopy, TLM = transmission light microscopy, SEM = scanning electron microscopy)

Cast-embedding technique

For the study of the microendolithic traces, a number of fresh bivalve shells (*Ostrea*, *Acantocardium*, *Callista*) as well as coral fragments (*Lophelia pertusa*) were attached to the frames. In order to reveal the endolithic boring traces, the following cast-embedding technique was applied (Fig. 5): after removing the organic matter via treatment with hydrogen peroxide and cleaning in an ultrasonic bath, the samples were placed in a vacuum chamber specifically designed for the cast embedding method (Struers Epovac). The chamber allows the infiltration of several samples with a high viscosity epoxy resin (Struers SpeciFix-20, or Araldite BY158) while the samples are held under vacuum. This way, a complete infiltration of the borings is assured and trapping of air is avoided. The latter method refines the time-

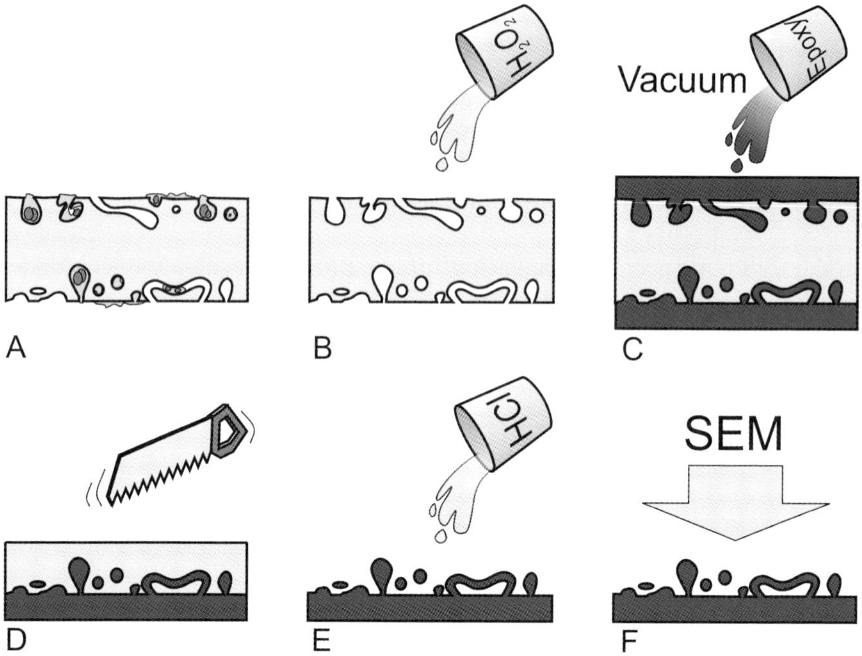

Fig. 5 Schematic sketch illustrating the cast-embedding technique: **A** The bored calcareous substrate with microendoliths and biofilms. **B** Removal of organic material by treatment with hydrogen peroxide and cleaning in an ultrasonic bath. **C** Infill with epoxy resin under vacuum conditions. **D** Formatting of blocks with a rock saw. **E** Removal of calcareous substrate via treatment with diluted hydrochloric acid. **F** SEM visualisation after sputter-coating with gold

consuming treatment with a sequence of different resin/acetone mixtures as applied by the established method after Golubic et al. (1970, 1983). After curing at 70°C, the samples were formatted and decalcified by a treatment with diluted HCl (~5%). The samples – now exhibiting the positive casts of the borings – were then sputter-coated with gold and finally analysed and photographed using a CamScan scanning electron microscope (SEM). In order to prevent delicate casts from collapsing and to help recognising the real penetration depths of the traces, some sample cross sections were only partially etched, allowing a visualisation of the traces in context of the shell matrix. Best results were obtained from the inner side of *Callista chione* shells, which were consequently preferentially chosen for the present evaluation. A pristine inner surface and the absence of any shell specific tubules or pores was beforehand assured by casting and scanning a dozen randomly chosen shells.

Visualisation of macroborings and calcareous epizoans

Macroborings and calcareous epizoans were photographed either by SEM or incipient light microscopy (ILM). For this purpose a CamScan SEM and Olympus SZX 12 with image analysis software (AnalySIS), respectively, were employed. Thin-sections were analysed and photographed with a Zeiss Axiophot transmission light microscope (TLM).

Visualisation of endoliths

The employed Iceland spars allowed the analysis of the endoliths in situ by transmission light microscopy (TLM). In this case, however, branching patterns and gallery directions of many endoliths may be biased by the crystal lattice of the substrate as reported by Golubic et al. (1975). The sugar-cube-sized micrite blocks as well as bivalve shells were prepared for light microscopy by dissolution of the substrate with Perenyi solution (4 parts 10% nitric acid, 3 parts 90% ethanol and 3 parts 0.5% chromic acid). This method assures a gentle isolation of the endolithic organisms which can than be studied and photographed with high-resolution light microscopes.

Quantitative analysis of bioerosion agents

The abundance of the various micro- and macrobioerosion traces in mollusc substrates and endoliths in Iceland spar was determined semi-quantitatively and is based on four abundance classes which were averaged by the number of samples investigated: 'very rare' (only one or very few specimens), 'rare' (few specimens), 'common' (many specimens but none dominant) and 'very common' (dominant). For the isolated endoliths from the shell material, only their presence was recorded. In addition, the shallow etching grooves left by foraminiferans and anomiid bivalves and the polychaete borings on the carbonate substrates were point-counted.

Quantitative analysis of calcareous epizoans

After retrieval, the PVC and carbonate plates were watered in freshwater followed by gentle treatment with diluted hydrogen peroxide (H_2O_2) in order to degrade all soft bodied tissue. After drying 48 hours at 70°C, all settled organisms were identified on family level (foraminiferans on species level) and point-counted prior to careful removal and weighing. Specimen numbers were then extrapolated to individuals per square metre after averaging the counts of the individual plates (counts on the carbonate plates were corrected for the area lost by the mounting nylon nut and washer).

Assessing bioerosion and carbonate accretion rates

In order to quantify bioerosion and carbonate accretion rates, roughened carbonate plates (an Upper Jurassic micritic limestone) and PVC plates, 10 x 10 x 1 cm in size, were prepared. The carbonate plates were sealed off with epoxy resin on all but the upper side in order to restrict bioerosion to a two-dimensional test surface allowing an easier and more accurate later assessment of bioerosion rates per surface area. All plates were dried at 70°C for 48 hours and weighed on a precision scale (±0.01 g for samples >100 g). Six plates of each type were fixed on the PVC experimental frames with nylon nuts and bolts.

After retrieval of the panels, the plates were removed and watered in freshwater followed by gentle treatment with diluted hydrogen peroxide (H_2O_2) in order to degrade soft bodied organisms and tissue. After drying 48 hours at 70°C, the plates were weighed (±0.01 g for samples >100 g). The encrusting calcareous organisms were then carefully removed and the carbonate accretion per plate was directly measured by weighing on a precision scale (±0.001 g for samples <100 g). The bioerosion per plate was calculated by subtracting the weight loss of the erosion plates from the carbonate accretion on the same plate. Rates were then averaged from all blocks at each depth station and extrapolated to the widely used unit grammes per square metre and year [$g/m^2/y$]. The surface area occupied by the mounting nut and washer was considered in the extrapolation.

In order to better distinguish carbonate accretion rates from bioerosion rates, the latter are given as negative values and a 'high rate' is consequently represented by more negative values.

Estimating carbonate accretion rates of foraminiferans

The annual carbonate production of benthic foraminiferans refers to the number and weight of tests produced per area and year. *Cibicides lobatulus* is the most abundant calcareous species in the sampling area representing mean 83.5% of the total population of carbonate-secreting benthic foraminiferans. To estimate the carbonate production rate of *Cibicides lobatulus*, a number of 250 specimens of different water depths and exposure time were weighed using the high-precision micro scale Sartorius M3P. The annual carbonate production of *Cibicides lobatulus* was calculated for 6, 12 and 24 months for each water depth using the average test weight of 0.0187 mg and the species density on the settlement plates.

3 The Kosterfjord study site

3.1 The northern Kosterfjord and the Säcken Reef site

The Kosterfjord is a submarine trench that parallels the coastline of SW Sweden to the east and is sheltered by numerous islands to the west. It deviates to NW, south of the Söstre Islands where it transits into the deep Norwegian Channel (Fig. 6A). The northern Kosterfjord is separated by the Säcken-sill from the Singlefjord (Fig. 6B). From the Säcken area the cold-water scleractinian coral *Lophelia pertusa* was sampled for the first time in Swedish waters (Wahrberg & Eliason 1926). With only around 85 metre water depth, the Säcken Reef site is one of the shallowest cold-water coral sites known to date (Jonsson et al. 2004). Here, the deepest experimental panels of the bathymetric transect were deployed. The shallower stations (~1 m, 7 m, 15 m, 30 m and 50 m) were located few kilometres to the north in a gully at the eastern face of the Singlefjord (Figs. 6B, 7). The second deployment site for the shallowest stations (~1 m at low tide) was located to the west of the TMBL in a sheltered and shallow lagoonal setting with soft, muddy ground.

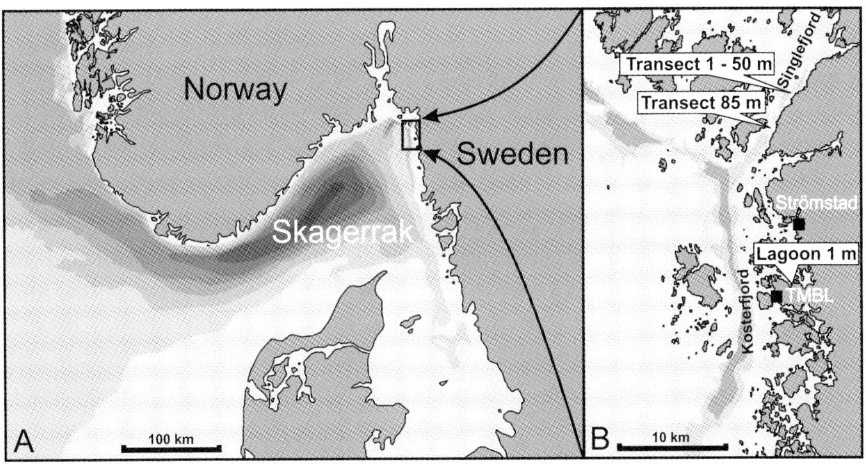

Fig. 6 A The geographic setting of the Kosterfjord in the NE Skagerrak. **B** Map of the northern Kosterfjord area with the location of the Tjärnö Marine Biological Laboratory (TMBL), and the deployment sites of the experimental panels (contours are given in 100 m intervals)

In the wider Kosterfjord area seven major and two minor *Lophelia* patches or mature reefs are known. The recently discovered Tisler Reef in neighbouring Norwegian waters (Lundälv & Jonsson 2003) is the largest structure being approximately 2 km in length. The entire area was glaciated until the Late Pleistocene. To date it is not known exactly when the evolution of the Kosterfjord reefs commenced. The oldest radiocarbon dates obtained from *Lophelia* skeletons from various sites off mid and southern Norway cluster around 8,700 to 8,600 years BP for the Early Holocene coral colonisation (Mikkelsen et al. 1982; Rokoengen & Østmo 1985).

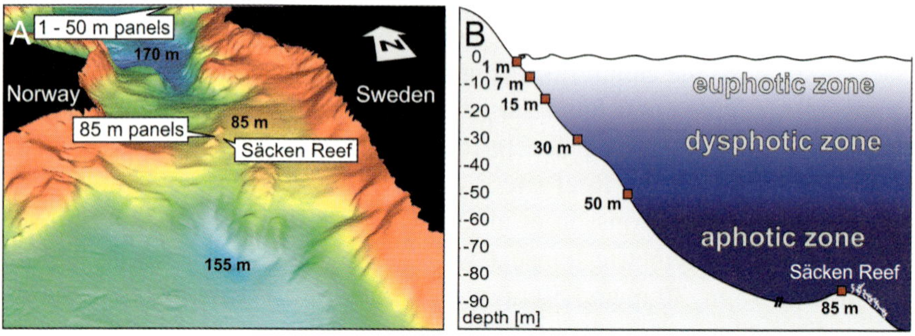

Fig. 7 A Detailed 3D bathymetry of the Säcken sill (generated from multibeam sonar data; modified from Jonsson et al. 2004) indicating the two *Lophelia* mound structures and the deployment sites of the experimental panels. **B** The bathymetric transect in the northern Kosterfjord area in relation to the photic zonation

The benthic substrate in the sound consists mostly of steep rocky cliffs or moraines with gravel and scattered stones. Two small mounds, in a NNE-SSW relationship, are situated on the sill. They are composed mostly of *Lophelia pertusa* rubble and densely packed silty clay. The live *Lophelia pertusa* occurrence forms two major patches, which are found on the SSW flanks of the two mounds (Fig. 7A). The patches are situated near the base of the mounds. The northern and southern patches cover areas of about 250 m^2 and 50 m^2, respectively. Most of the colonies are small and grow as discrete, often spherical colonies with lumps of dead coral, coral rubble, gravel, cobble and sediment in between. At the present time, a total of 307 species have been identified associated with *Lophelia pertusa* at the Säcken site (Lundälv pers. comm.).

3.2 Oceanography and hydrology

General patterns

The Kosterfjord lies in the northeastern part of the Skagerrak which is the major gateway between the North Atlantic and the Baltic Sea. The overall oceanographic regime is driven by an estuarine circulation pattern. Incoming dense, saline (30-35 psu) and oxygenated oceanic water underflows the more brackish (8-30 psu) surface-water outflow of the Baltic Sea (Svansson 1975). Off the northern Swedish west coast, in the Kosterfjord area, the surface water circulates with a pre-dominating northern heading consisting of the Northern Jutland Current, which continues as the Norwegian Coastal Current along the Norwegian coastline. In addition, more brackish water is added from the Baltic Current, entering the Skagerrak through the Kattegat and the Sound to the south (Svansson 1975). This surface circulation is compensated by a deep counter current that brings the saline Atlantic water through the up to 700 m deep Norwegian Trench (Fig. 6A) into the Skagerrak and the Kattegat (Dahl 1978). Tidal sea-level fluctuations are below 0.5 m but those related to atmospheric pressure changes may exceed 1 m.

Seasonal fluctuations

The general oceanographic watermass stratification of the Skagerrak is also clearly expressed in the Kosterfjord area. The sea surface temperature (SST) is subject to strong seasonal fluctuations, ranging from below freezing (sea ice) in the two coldest months February and March to 17-20°C in the warmest months July and August (Fig. 8A-B). This annual fluctuation is also present in deeper waters, but with lower amplitude and a certain time delay – in 100 m water depth for instance, temperatures vary between 4 and 10°C and the amplitudes are delayed by about three to four months (Fig. 8A). The salinity plots draw a similar picture (Fig. 8C-D): The strongest variations, which are due to varying degrees of freshwater input, are exhibited by the surface waters, where they range from a minimum of as low as 8.0 psu to 30.9 psu. The lowest salinities are present between May and August. In 100 m water depth there is only a low annual salinity variation between 33.6 psu and 35.0 psu with a mean of 34.5 psu. Temperature- and salinity-depth-plots (Fig. 8B and D) indicate a thermocline and halocline, respectively, at around 20 to 40 m water depth, separating the open marine Atlantic waters from the brackish outflow of the Baltic Sea for most of the year. Another boundary layer is encountered in only a few decimetres to metres water depth, where a brackish to near-freshwater layer is added by the local river runoff. This stratification is most prominent in the coldest (February and March) and warmest (July and August) months in respect to the SST. The reduced salinity of the surface water facilitates the formation of sea ice during harsh winters. The oxygen level of the surface water ranges from 4.2 to 9.5 ml/l with the highest values in springtime and the lowest in autumn (reverse proportional to the water temperature). Analogously to temperature and salinity, the fluctuation is smaller at 100 m water depth, where the values plot at 4.8 to 6.7 ml/l.

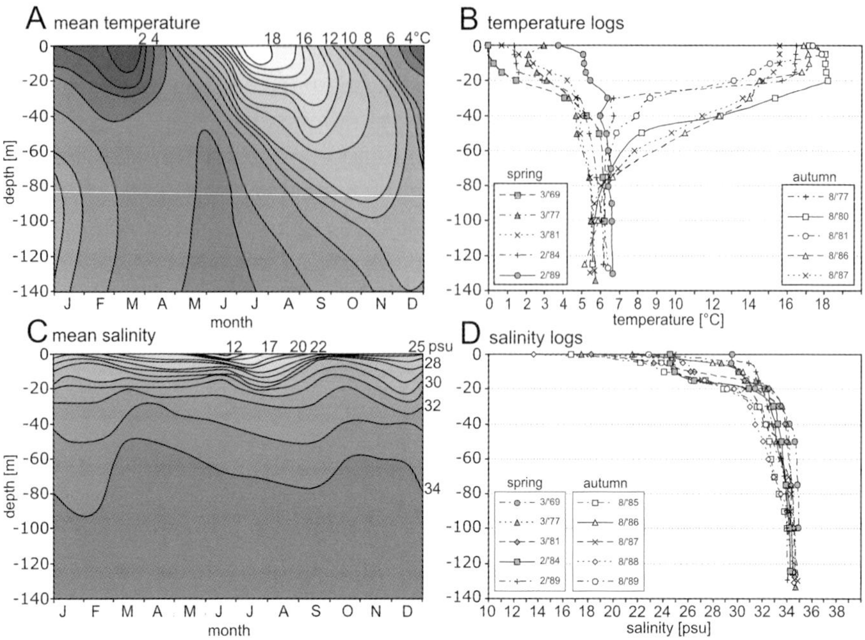

Fig. 8 Temperature and salinity variations in the northern Kosterfjord area compiled from over 700 measurements logged just south of the sample sites (59°00'4" N / 11°06'8" E) between 1967 and 1990. **A** Annual temperature fluctuation (based on monthly means) with a declining amplitude and typical time delay towards deeper waters. **B** Five representative temperature logs for springtime and autumn, respectively, exhibiting a main thermocline in 20 to 40 m water depth. **C** Annual salinity fluctuation (based on monthly means) indicating a major freshwater influence in the shallow waters. **D** Five representative salinity logs for springtime and autumn showing a halocline at around 20 m water depth

Short-term fluctuations

The two lander deployments carried out near the *Lophelia* patches at the Säcken Reef site at ca. 85 m water depth recorded temperature, salinity, current direction, current speed and turbidity in one hour intervals (Table 6). The temperature at this depth is subject to only minor short-term fluctuations of less than one degree Celsius. The mean autumn temperature plots at 7.9°C while the springtime mean was about two degrees lower at 6.2°C. The autumn salinities ranged from 34.4-35.1 psu with a mean of 34.8 psu and short-term fluctuations in the range of 0.4 psu. The mean spring salinity was a little lower (34.6 psu), with maximum values of around 34.8 psu. The turbidity was relatively constant at around 0.5 ntu. Short-term fluctuation of all parameters except turbidity was lower in springtime than in autumn. The current speed occasionally reached close to 25 cm/s, with a mean of 6 cm/s in autumn and 4.5 cm/s in springtime. The current direction measurements are influenced by the small- to medium-scale morphology of the reef mounds and are thus not representative for the whole reef site. Nevertheless, it is worth noting that the direction shifted in a cyclic pattern between two main directions. This was especially pronounced in springtime, where the direction shifted between 100° and 330° heading, which are the directions with the highest current velocities.

Table 6 Hydrographic parameters at the Säcken Reef site in 85 m water depth recorded in one hour intervals by a lander during September-October 2000 and February-April 2001

	Autumn 2000 (Sept 6 - Oct 5, 2000)				**Spring 2001 (Feb 22 - Apr 4, 2001)**			
	M	Min	Max	SD	M	Min	Max	SD
Temperature [°C]	7.9	7.2	8.7	0.2	6.2	5.6	7.1	0.4
Salinity [psu]	34.8	34.4	35.1	0.2	34.6	34.3	34.8	0.2
Turbidity [ntu]	0.5	0.3	1.0	0.1	0.6	0.3	4.6	0.2
Current velocity [cm/s]	6.0	0.0	24.0	4.4	4.5	0.0	21.0	2.9
Current direction [°]	203.0				229.0			

3.3 The photic zonation

The concept of the photic zonation

The light availability in relation to bathymetry is one of the main factors influencing the distribution of microendoliths and determining the composition of the borer communities. The scheme of the photic zonation is applied in the present study. Liebau (1984) and Glaub (1994) distinguish a shallow- and deep-euphotic zone, a dysphotic zone and an aphotic zone. The base of the euphotic zone is defined by the depth where the light intensity declines to 1% of the surface illumination. The base of the dysphotic zone corresponds to the photic limit, which is the depth at which photosynthesis balances respiration and below which photoautotrophic algae do not exist (~0.01% of the surface illumination).

Defining the illumination status

Due to a lack of suitable literature data, the photosynthetically active radiation (PAR) was directly measured under various weather conditions and seasons (always around noon), in order to determine the depth of the boundaries of the photic zonation in the study area. The surface illumination ranged from ~600 to 2,400 μmol photons m^{-2} s^{-1} above and ~200 to 1,400 μmol photons m^{-2} s^{-1} just below the water surface, and light levels decreased exponentially toward deeper waters (Fig. 9A). The base of the euphotic zone was directly measured in 10-25 m water depth while the base of the dysphotic zone was estimated by plotting the data and a regression line on a logarithmic scale (Fig. 9B), yielding a range of 30-60 m water depth (for the calculation of the regression line, the uppermost water column was excluded because of its significantly higher turbidity). Even though measurements were not taken during the summer period, where we can expect a somewhat deeper light penetration, this dataset clearly reflects a highly condensed photic zonation as compared to tropical oligotrophic settings, where most of the previous experimental studies on bioerosion have been carried out.

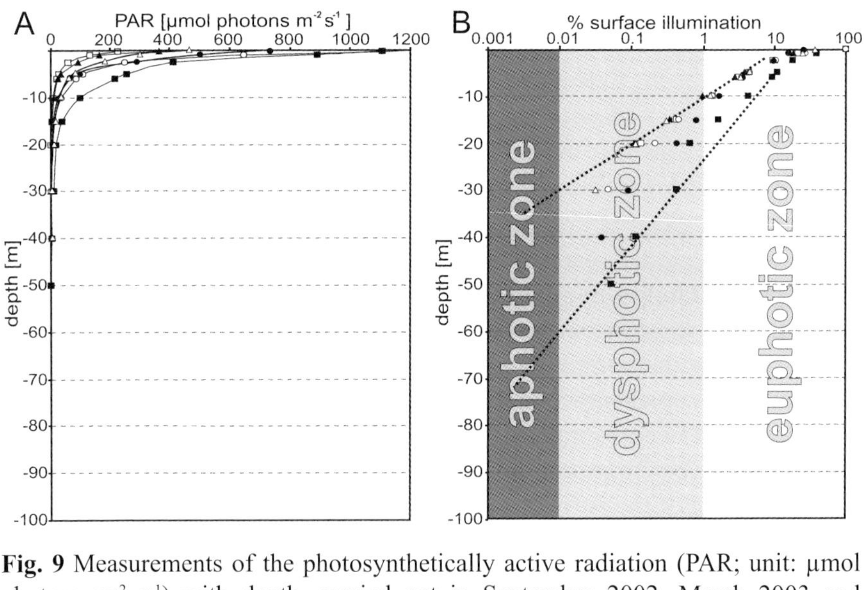

Fig. 9 Measurements of the photosynthetically active radiation (PAR; unit: μmol photons m^{-2} s^{-1}) with depth, carried out in September 2002, March 2003 and October 2003. **A** The illumination exponentially decreases toward deeper waters. **B** Logarithmic plot with regression lines indicating a boundary to the dysphotic zone (1% surface illumination) in 10 to 25 m water depth and the upper limit of the aphotic zone (~0.01% surface illumination) in a range of 30 to 60 m water depth

4. Bioerosion patterns

Traces of microborers (gallery diameter <100 µm; Sect. 4.1), grazers and macroborers (gallery diameter >100 µm; Sect. 4.2) were primarily investigated in the experimental substrates, especially in *Callista* bivalve shells and micrite blocks. In the case of microbioerosion, the study of the boring traces was coupled to a biological identification of the producing microendoliths, both in situ in the Iceland spars and transparent bivalve shells as well as isolated from opaque shells (carried out by Marcos Gektidis). In addition to the experimental substrates, background sample material was analysed from the two end members of the bathymetric transect, namely from skeletons of the cold-water coral *Lophelia pertusa* and associated bivalves sampled at the Säcken Reef site in 85 m water depth (Sect. 4.3) and from the shallow-water gastropod *Littorina littorea* collected at the shoreline (Sect. 4.4). By choosing a sessile organism in the first and a bathymetrically highly restricted species in the second case, any bathymetric bias due to transport of the shell material was minimised. While the experimental substrates document the early to intermediate stages of endolithic colonisation, the background samples record advanced taphonomical stages of shell degradation, allowing a more complete picture of the bioerosion agents and traces present in the study area.

4.1 The microbioerosion inventory

In the following account, a detailed inventory of the microboring ichnotaxa as well as their known or assumed producers is given. Since for most ichnotaxa the original diagnosis was introduced to the literature in German (Radtke 1991; Schmidt 1992; Glaub 1994) and in one case in Spanish language (Mayoral 1988), an English translation is given enabling access for a broader readership. Traces which require further investigation are presented in informal nomenclature (e.g., '*Fascichnus*-form 1').

The inventory is chiefly based on the microendoliths recorded in the experimental substrates and their traces studied by SEM of epoxy resin casts. Only the bryozoan traces alongside '*Orthogonum*-form 3', 'Foraminiferan-form 3' and 'Problematic-form 1' were exclusively encountered in the aphotic background sample material (see Sect. 4.3) and *Fascichnus rogus* only in the shallow-euphotic material (see Sect. 4.4). In the experimental substrates, a total of 21 different traces were produced by cyanobacteria (7), green algae (4), fungi (6) and of unidentified origins (4) (Fig. 10). The investigation of the Iceland spar yields 12 different

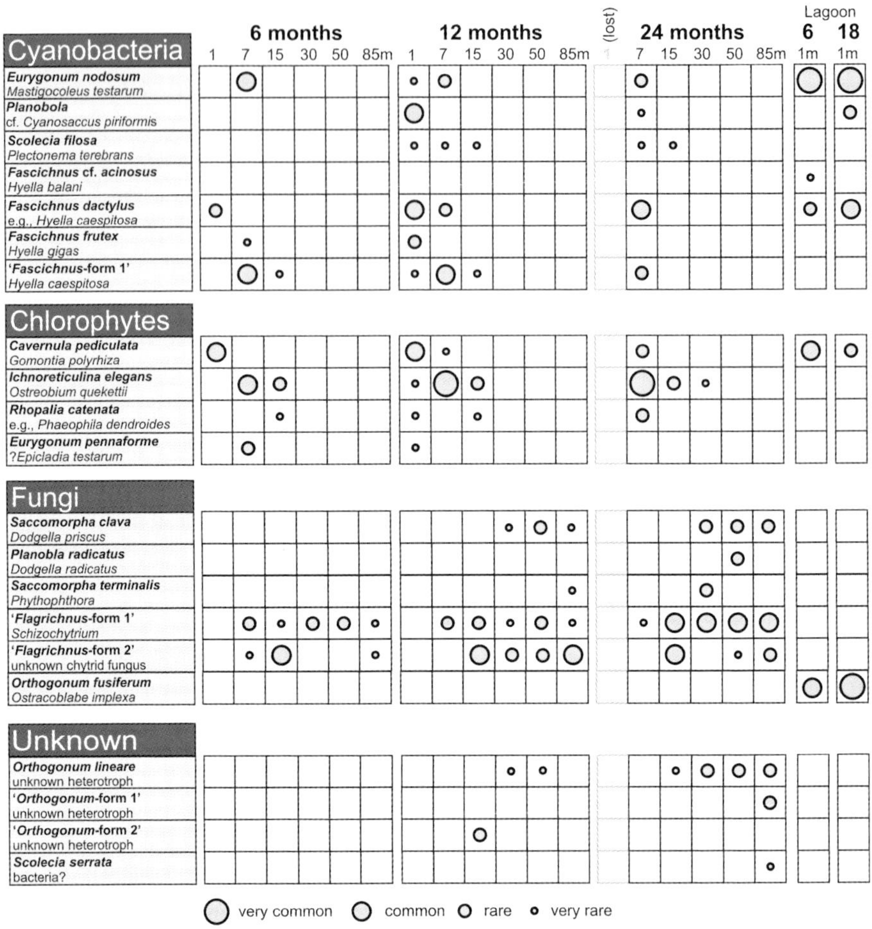

Fig. 10 Semi-quantitative analysis of the microendolithic traces, based on SEM analysis of resin casts from *Callista chione* shells that were deployed along a bathymetric transect for 6, 12 and 24 months

endolithic taxa comprising cyanobacteria (5), green algae (5), fungi (1) and one unknown form. The analysis of the shell material yields 17 different endolithic taxa consisting of cyanobacteria (8), green algae (5), fungi (3) and one unknown form (Fig. 11). Due to the delicate nature of microendolithic fungi, some with hyphal diameters around 1 μm, their isolation was only sporadically successful. In the semi-quantitative analysis, this is reflected in a biased rare occurrence, when compared to the analysis of the boring casts. The total number of microboring ichnotaxa encountered in all investigated substrates sums up to 26.

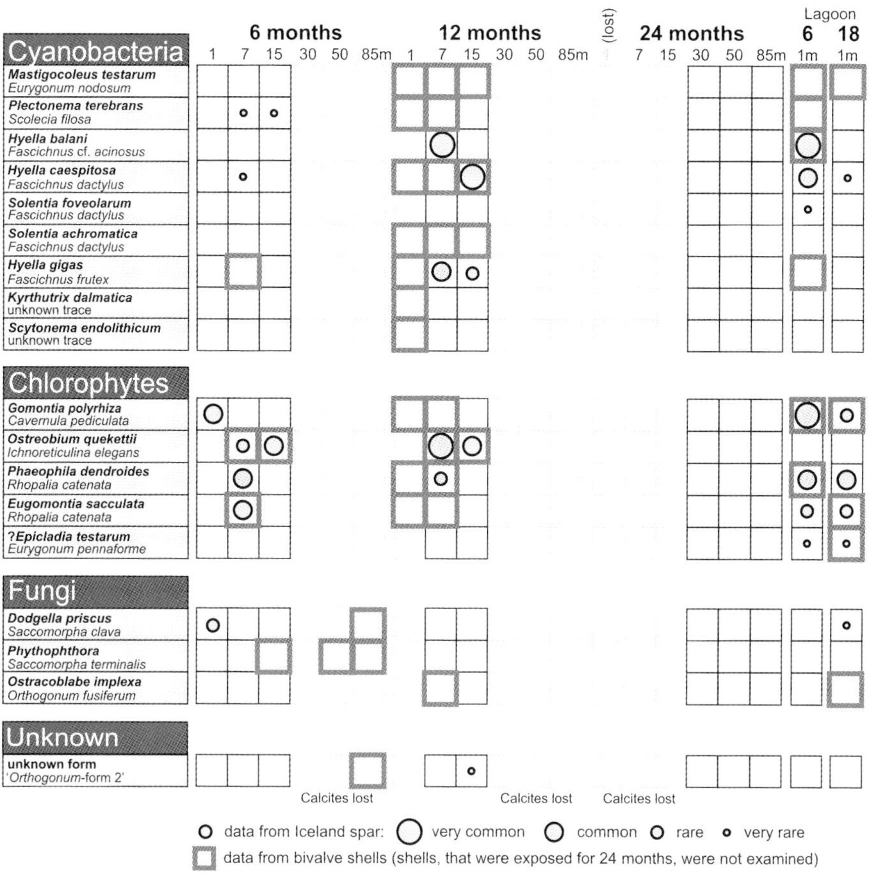

Fig. 11 Semi-quantitative analysis of the microendolith distribution as observed by transmission light microscopy in Iceland spar and isolated from *Callista* shells (qualitative) that were deployed along a bathymetric transect for 6, 12 and 24 months

Cyanobacteria

Ichnotaxon ***Eurygonum nodosum*** Schmidt, 1992
Figs. 12A-B, 31F
Trace maker ***Mastigocoleus testarum*** Lagerheim, 1886
Figs. 12C, 20H

Ichnospecies diagnosis: Horizontal, three-dimensional, multi-axial galleries with dichotomous bifurcations, terminal swellings and lateral appendices (after Schmidt 1992).

Description: Two morphotypes are developed, the first one (Fig. 12A) forming radiating colonies up to 1 mm in diameter, oriented closely parallel to the substrate surface. From a central area of the colony, a network of galleries (ø 5-8 μm) radiate and repeatedly bifurcate in an uni- or a bilateral mode in angles between 45° and 90°. The galleries may bear slight terminal swellings (up to 10 μm). Along the galleries many characteristic lateral swellings or distinct nodular appendices (ø 7-10 μm) are developed. Occasionally, those appendices cluster in larger numbers. The second morphotype (Fig. 12B) is distinguished by a more three dimensional network, fewer lateral appendices and indistinct colony boundaries.

Trace maker: The cyanobacterium *Mastigocoleus testarum* is characterised by trichomes with irregular branching and diameters of the thalli between 6 and 10 μm. The terminations of the trichomes are sometimes whip-shaped. The thalli exhibit diagnostic intercalar, terminal or lateral heterocysts. *Mastigocoleus testarum* propagates by hormogonia (e.g., Gektidis 1997b).

Distribution: Experimental substrates: 1-7 m stations; background samples: 1 m

Remarks: The galleries in our material have a relatively small diameter (5-8 μm) as compared to the type material (9-11 μm). However, Glaub (1994) gives a range of 7-12 μm for this trace and 6-10 μm for the presumed trace maker *Mastigocoleus testarum*. The, in parts, relatively large number of nodular appendices as shown by the first morphotype is a feature rarely observed in this ichnospecies. This feature and the general branching pattern (showing some similarity with *Eurygonum pennaforme*) of this morphotype could alternatively indicate a chlorophyte alga as potential trace maker. Such specimens closely resemble the 'Problematic algal form A' as depicted by Budd & Perkins (1980: fig. 7A).

Ichnotaxon ***Planobola*** Schmidt, 1992
Figs. 12D-E, 31G
Trace maker cf. ***Cyanosaccus piriformis*** Lukas & Golubic, 1981
Gomontia polyrhiza (Lagerheim) Bornet & Flahault, 1888
Fig. 15B

Ichnogenus diagnosis: Spheroid to bulboid boring system with latitudinal contact to the substrate surface (after Schmidt 1992).

Description: Globular cavities, 10-30 μm in diameter connected to the substrate surface by a single, thick stalk.

Trace maker: This ichnogenus addresses featureless globular traces, which may resemble initial borings of various trace makers such as the green alga *Gomontia polyrhiza* and various cyanobacteria, as well as unicellular or globular multicellular cyanobacteria such as *Cyanosaccus piriformis*.

Distribution: Experimental substrates: 1-7 m stations; background samples: 1 m.

Remarks: Due to the lack of features, no assignation of the present material to a specific ichnospecies of *Planobola* is undertaken.

Ichnotaxon ***Scolecia filosa*** Radtke, 1991
Figs. 12F-G, 31H
Trace maker ***Plectonema terebrans*** Bornet & Flahault, 1889
Fig. 12H

Ichnospecies diagnosis: Long, thin, often worm-like bent or curled galleries, which are occasionally developing a network. Bifurcations are rare (emended after Radtke 1991).

Description: Thin (1-2 μm), only occasionally bifurcating galleries of uniform diameter, commonly found collapsed to the cast surface.

Trace maker: Plectonema terebrans is a small endolithic species (1-1.5 μm). Trichomes are curved, have false branchings and do not exhibit heterocysts. This cyanobacterium propagates by hormogonia (Bornet & Flahault 1889).

Distribution: Experimental substrates: 1-15 m stations; background samples: 1 m.

Remarks: This ichnotaxon is difficult to interpret since it is (1) hard to tell apart from fungal hyphae of for instance *Saccomorpha* or the deeply penetrating galleries of the two '*Flagrichnus*-forms', and (2) it is commonly found collapsed to the cast surface making the original three-dimensional architecture difficult to identify. Hence, the trace is only apparently oriented closely parallel to the substrate surface and the phrase "close to the substrate surface" is consequently omitted in the emended diagnosis here.

Ichnotaxon ***Fascichnus* cf. *acinosus*** (Glaub, 1994)
Fig. 13A
Trace maker ***Hyella balani*** Lehmann, 1903
Fig. 13B

Ichnospecies diagnosis: Groups of galleries, penetrating the substrate in a primary perpendicular manner from a central area on the substrate surface, showing proximal fused and distally separated galleries (after Glaub 1994).

Description: From a central area on the surface, the colony is radiating laterally and/or perpendicular into the substrate. Individual galleries are proximally fused and only separated at their thinner (~5 µm) distal parts. Occasionally, cell-like constrictions are developed.

Trace maker: Hyella balani was originally introduced by Lehmann (1903) and has been redefined by Le Campion-Alsumard & Golubic (1985). The latter authors describe four morphologically distinct ecotypes. 'Status typicus', which is the only type found in our experiment, is characterised by an apical cell-length of 11.5-18.9 µm and cell-width of 5.8-7.4 µm.

Distribution: Experimental substrates: 1 m station.

Remarks: The specimen only partly resemble the material and original diagnosis given by Glaub (1994), since the colonies are spreading in a more lateral fashion and individual galleries are more distinct. The genus name *Fascichnus* was most recently introduced by Radtke & Golubic (2005) replacing the former *Fasciculus* due to priority rights (nomen nudum).

Ichnotaxon ***Fascichnus dactylus*** (Radtke, 1991)
Figs. 13C-E, 31A
Trace maker ***Hyella caespitosa*** Bornet & Flahault, 1889
Fig. 13F
Solentia foveolarum Ercegovic, 1927, emend. 1932
Fig. 13G
Solentia achromatica Ercegovic, 1932
Fig. 13H

Ichnospecies diagnosis: A bundle of uniformly thick, separate galleries, sometimes with terminal club-shaped swellings, penetrates the substrate from a central area mainly in a perpendicular or angled direction (after Radtke 1991).

Description: The trace is characterised by either a radiating bundle or larger carpet of up to 150 µm long galleries, 3-8 µm in diameter, which is penetrating into the substrate. The individual galleries are uniformly thick or may slightly thicken towards their distal ends. Bifurcations are rare and only present near the base of the galleries. There is a wide range from deeply penetrating thin forms to short and thick variants (Fig. 13E).

Trace maker: This ichnospecies is produced in akin morphology by a number of endolithic cyanobacteria, three of which were identified in the Iceland spar and bivalve shells: (1) *Hyella caespitosa* is a collective species sensu Geitler (1932). In the present study all *Hyella* colonies with cell widths between 4 and 10 µm and thalli up to 60 µm in length were subsumed under this species unless a distinct morphology led to a specific identification. (2) *Solentia foveolarum* exhibits thalli up to 60 µm wide. Individual cells are 5-20 µm wide and up to 40 µm long. The distance between apical and vegetative cells of the colony varies from 10-400 µm (Ercegovic 1932). (3) The cells of *Solentia achromatica* are 7-16 µm wide and 15-90 µm long. The distance between apical and vegetative cells in this case ranges from 12-28 µm (Ercegovic 1932). All three species propagate by baeocytes.

Distribution: Experimental substrates: 1-7 m stations; background samples: 1 m.

Remarks: Radtke & Golubic (2005) replaced the former *Fasciculus* (nomen nudum) by the new genus name *Fascichnus*.

Ichnotaxon ***Fascichnus frutex*** (Radtke, 1991)
Figs. 14A-C, 31B
Trace maker ***Hyella gigas*** Lukas & Golubic, 1983
Fig. 14D

Ichnospecies diagnosis: From a central area, thick and dichotomous ramifying galleries with round, blunt ends radiate into the substrate (after Radtke 1991).

Description: In the present material, two different morphotypes of *Fascichnus frutex* can be distinguished. The first one (Fig. 14A-B) comprises mainly uniformly thick (ø 10-15 µm) galleries, which frequently bifurcate and bear distinct fine constrictions. The second morphotype (Fig. 14C) is characterised by a bundle up to one millimetre in diameter, of distally thickened galleries, 10-30 µm in diameter, radiating from a central area into the substrate. The individual tunnels exhibit a club-shaped appearance and sometimes bifurcate in their distal part.

Trace maker: Hyella gigas is characterised by large cells (up to 28 µm in length and 22 µm wide; apical cells up to 50 µm) forming thalli. *Hyella gigas* propagates by baeocytes (Lukas & Golubic 1983).

Distribution: Experimental substrates: 1-7 m stations; background samples: 1 m

Remarks: Radtke & Golubic (2005) replaced the former *Fasciculus* (nomen nudum) by the new genus name *Fascichnus*.

Ichnotaxon ***Fascichnus rogus*** (Bundschuh & Balog, 2000)
Fig. 31C-D
Trace maker **cf. *Hyella racemus*** Al-Thukair, Golubic & Rosen, 1994

Ichnospecies diagnosis: Fascichnus rogus has the overall shape of a raspberry or cauliflower. The cross-section of the boring appears fan-shaped with the proximal tubes arranged tightly together. The individual tubes are distinguishable only by their separated, distal ends (Bundschuh & Balog 2000).

Description: The trace is characterised by raspberry-shaped aggregates, 60-120 µm in diameter, composed of several hundred distinct or fused 'cells' measuring in the range of 5-8 µm in diameter each.

Trace maker: The Silurian type material was compared by Bundschuh & Balog (2000) to the modern *Hyella racemus* (Al-Thukair et al. 1994) and possibly also *Hyella conferta* as described by Al-Thukair & Golubic (1991). Both cyanobacteria are, however, only known from shallow tropical settings and were not encountered in the Kosterfjord material.

Distribution: Background samples: 1 m.

Remarks: Radtke & Golubic (2005) replaced the former *Fasciculus* (nomen nudum) by the new genus name *Fascichnus*. The present material extends the known stratigraphic range (Silurian-Jurassic; Bundschuh & Balog 2000) to the Recent.

Ichnotaxon '***Fascichnus*-form 1**'
Figs. 14E-F, 31E
Trace maker ***Hyella caespitosa*** Bornet & Flahault, 1889
Fig. 14G

Description: The colonies comprise repeatedly bifurcating, often curved and distinctly flattened galleries, 8-10 μm in diameter, radiating from a central area closely parallel to the substrate surface. In the centre of the colonies, galleries may be fused to large irregular patches from which short and rarely branching galleries penetrate deeper into the substrate (Fig. 14F).

Trace maker: The traces lie within the morphological variability of *Hyella caespitosa* which is described under the ichnospecies *Fascichnus dactylus*.

Distribution: Experimental substrates: 1-15 m stations; background samples: 1 m.

Remarks: Glaub (2004: fig. 2f) depicts a similar trace but of smaller size.

Ichnotaxon **unknown**
Trace maker ***Kyrthutrix dalmatica*** Ercegovic, 1929
Fig. 14H

Trace maker: Heterocystous cyanobacterium with 15-20 μm wide and up to 400 μm long filaments, comprising single rows of cells which are sometimes u-shaped, sometimes parallel oriented. Cells are barrel-shaped, 4-7 μm wide and 5-9 μm long. Heterocysts are intercalar and approximately 9 μm wide. *K. dalmatica* propagates by 25-40 μm long hormogonia (Ercegovic 1929).

Distribution: Experimental substrates: 1 m station.

Ichnotaxon **unknown**
Trace maker ***Scytonema endolithicum*** Ercegovic, 1932

Trace maker: Scytonema endolithicum is a heterocystous cyanobacterium with 14-20 μm wide filaments. The trichomes are 8-10 μm wide and the heterocysts are intercalar and approximately 8-10 μm wide. The colonies exhibit an epi- as well as an endolithic part (Ercegovic 1932).

Distribution: Experimental substrates: 1 m station.

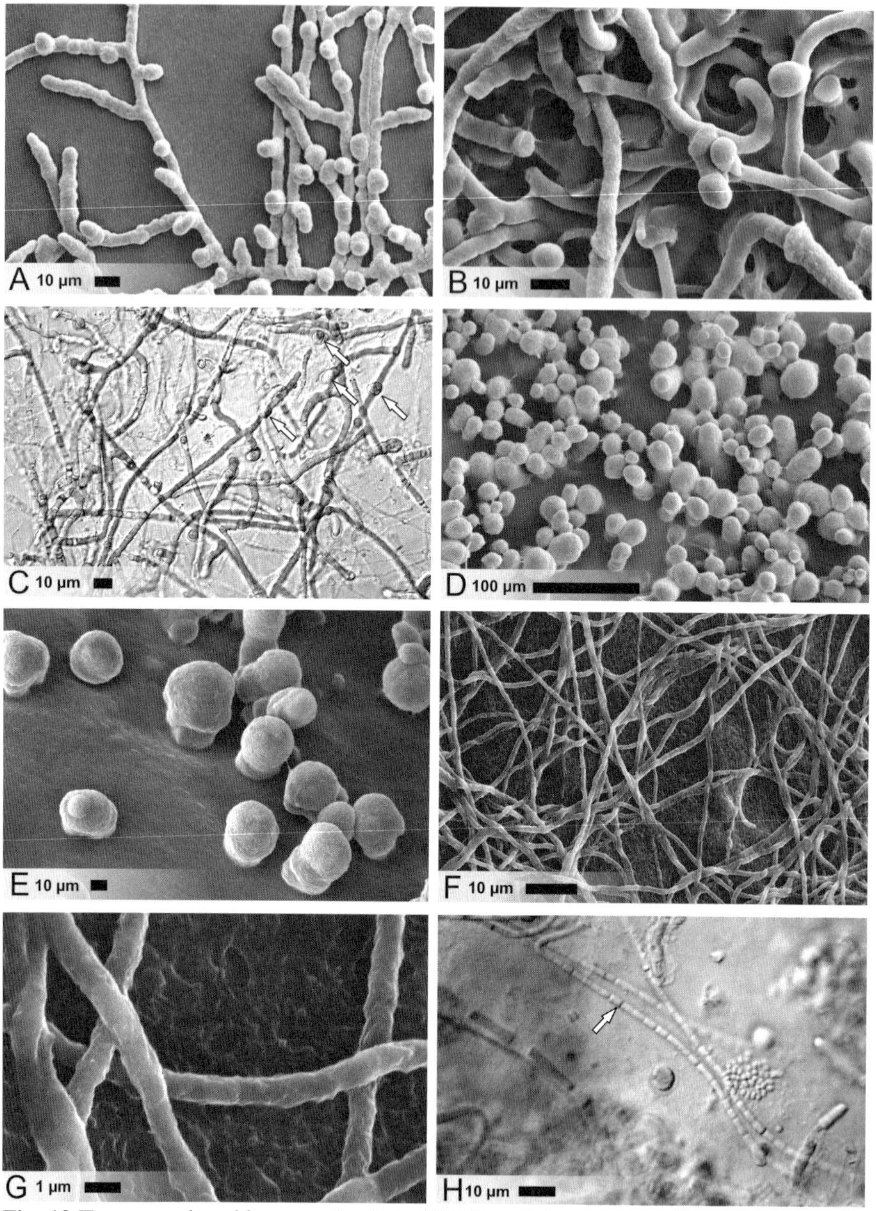

Fig. 12 Traces produced by cyanobacteria (SEM) and the corresponding microendoliths (TLM): **A** First morphotype of *Eurygonum nodosum* (7 m; 6 months). **B** Second morphotype of *E. nodosum* (1 m L; 18 months). **C** *Mastigocoleus testarum* with diagnostic heterocysts (arrows; 1 m L; 18 months; I). **D** *Planobola* (1 m; 12 months). **E** *Planobola*, illustrating the morphological affinity to juvenile *Cavernula pediculata* (1 m; 12 months). **F** The filamentous *Scolecia filosa* (15 m; 24 months). **G** Close-up of uniformly thin galleries of *S. filosa* (15 m; 24 months). **H** *Plectonema terebrans* (arrow; 1 m; 24 months; I)

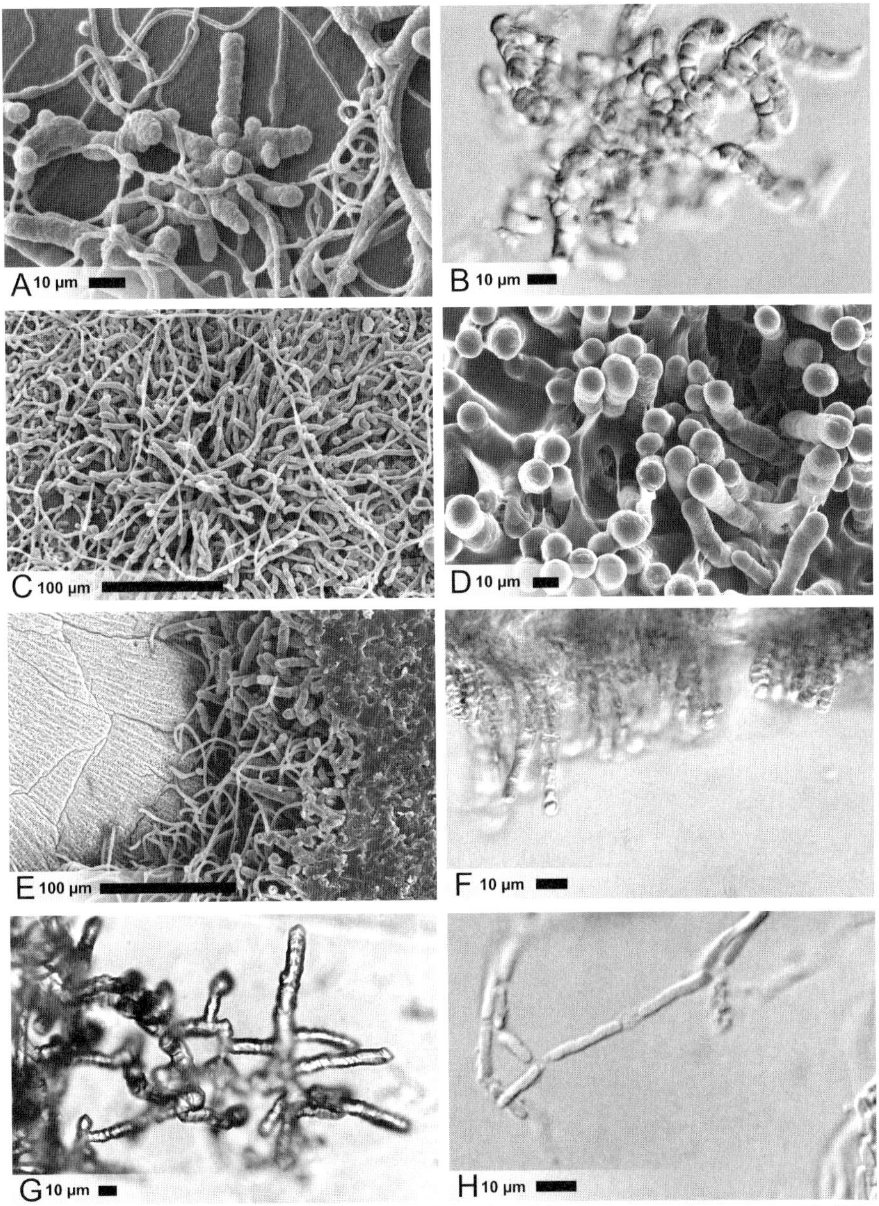

Fig. 13 Traces produced by cyanobacteria (SEM) and the corresponding microendoliths (TLM): **A** *Fascichnus* cf. *acinosus* partly covered by *Orthogonum fusiferum* (1 m L; 6 months). **B** *Hyella balani* (1 m L; 6 months; I). **C** Typical radiating colony of *Fascichnus dactylus* (7 m; 12 months). **D** *F. dactylus* with short and thick galleries (1 m; 6 months). **E** Both morphotypes of *Fascichnus dactylus* together in a partially etched section (1 m; 12 months). **F** *Hyella caespitosa* (1 m; 24 months; I). **G** *Solentia foveolarum* (1 m L; 6 months; IS). **H** *Solentia achromatica* (1 m L; 6 months; I)

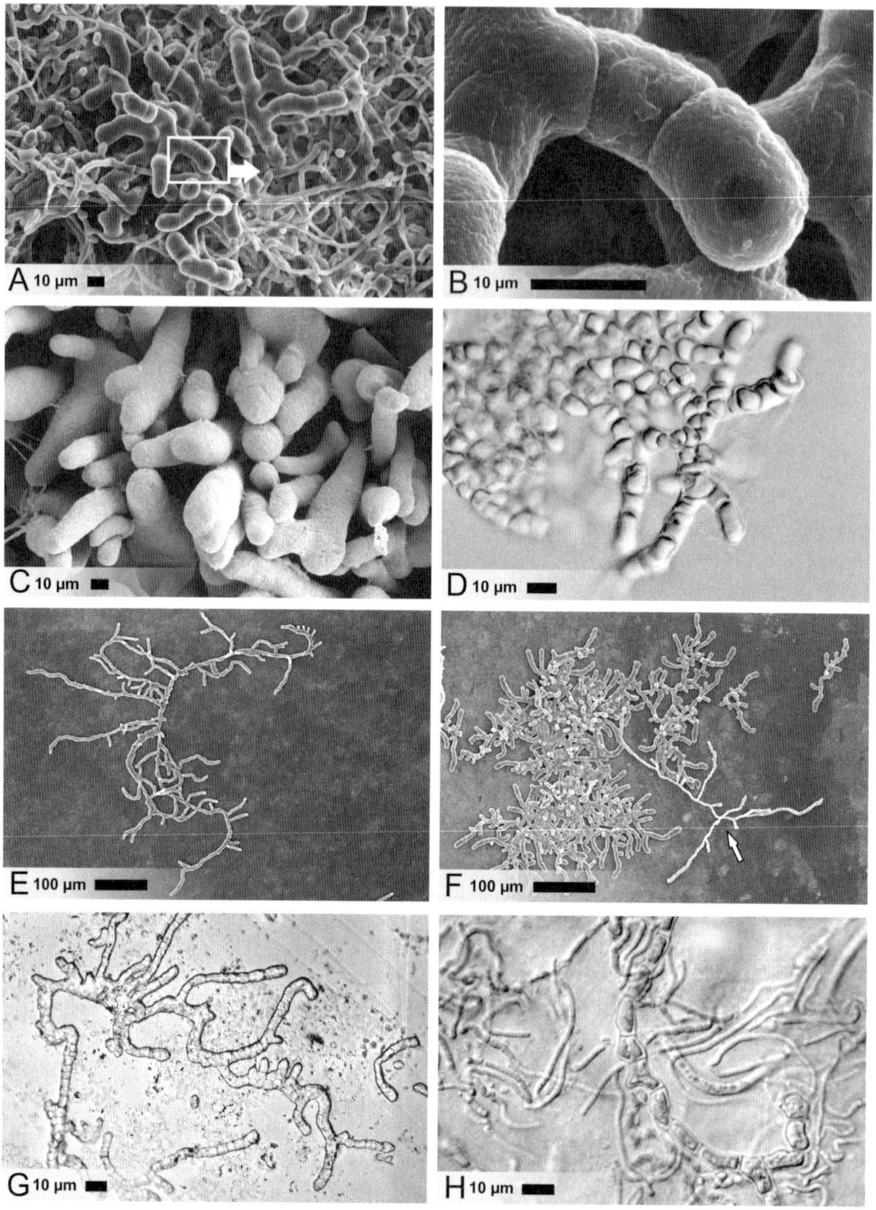

Fig. 14 Traces produced by cyanobacteria (SEM) and the corresponding microendoliths (TLM): **A** First morphotype of *Fascichnus frutex* (7 m; 24 months). **B** Detail of *F. frutex* with distinct constrictions. **C** Second morphotype of *F. frutex* (1 m; 12 months). **D** *Hyella gigas* (1 m L; 6 months; I). **E** Widely extending colony of '*Fascichnus*-form 1' (7 m; 6 months). **F** Dense colony of '*Fascichnus*-form 1' together with *Eurygonum pennaforme* (arrow; 7 m; 6 months). **G** *Hyella caespitosa* (15 m; 12 months; IS). **H** *Kyrthutrix dalmatica* overlain by a thick chlorophyte boring (1 m; 24 months; I)

Chlorophytes

Ichnotaxon ***Cavernula pediculata*** Radtke, 1991
Figs. 15A+C, 32A-B
Trace maker ***Gomontia polyrhiza*** (Lagerheim) Bornet & Flahault, 1888
Fig. 15B

Ichnospecies diagnosis: Solitary, large, sack-, bag-, to pear-shaped cavities, approximately perpendicular to the substrate surface, mostly connected to the latter by a number of rhizoidal appendages (after Radtke 1991).

Description: The traces are 15-40 µm wide and 20-100 µm deep, spherical to bag-shaped cavities, connected to the surface by a number of mostly short (max. 30 µm) rhizoidal appendages, thinning and occasionally bifurcating towards the substrate surface. The intra-specific morphological range includes wide and flat specimens, where radial rhizoidal appendages are longer and enter the substrate in more shallow angles.

Trace maker: The traces are produced by the single-celled *Codiolum*-stage of *Gomontia polyrhiza* and have a cell-width of 25-60 µm and cell-lengths of 60-100 µm. Originally described by Lagerheim (1885) as *Codiolum polyrhizum*, it was recognised as part of the chlorophyte alga *Gomontia polyrhiza* by Bornet & Flahault (1889). It was later again considered as individual species sensu Lagerheim by Kornmann (1959) but is currently regarded as synonym of *Gomontia polyrhiza*.

Distribution: Experimental substrates: 1-7 m stations; background samples: 1 m.

Remarks: The morphological range found in the present specimens corresponds well to the material documented by Kornmann (1959: fig. 7). Immature specimens, which exhibit only few, thick rhizoidal appendages, are indistinguishable from *Cavernula coccidia* Glaub, 1994, or may show some morphological affinity to the ichnogenus *Planobola* (cf. Fig. 12E).

Ichnotaxon ***Ichnoreticulina elegans*** (Radtke, 1991)
Figs. 15D+F+H, 32D
Trace maker ***Ostreobium quekettii*** Bornet & Flahault, 1889
Fig. 15E+G

Ichnospecies diagnosis: Thin, straight main galleries close to the substrate surface with many dichotomous branching, slightly thinner, zigzag-shaped galleries which may exhibit antler-like terminal swellings. The galleries often form a dense network (after Radtke 1991).

Description: From a straight or winding, circular to flattened main gallery ~5 µm in diameter, second order galleries emerge mostly at right angles and show a characteristical zigzag pattern. Occasionally, collar-like swellings appear along the galleries. Initial colonies are oriented close to the substrate surface with galleries rarely crossing each other while mature colonies form dense networks with several tiers. Quite frequently, large dendritic cavities ~100-150 µm in diameter (Fig. 15F), with an irregularly-shaped central main part, which occasionally shows a distinct sack-shaped protrusion, are encountered. From these cavities, numerous tapering galleries radiate and either end blindly or bifurcate repeatedly and merge into common main galleries. A previously unknown feature of *Ichnoreticulina elegans* was observed, namely up to 300 µm long tapering whip-shaped branches with a characteristic rough pipe-cleaner-like texture (Fig. 15H).

Trace maker: The well-studied siphonal alga *Ostreobium quekettii* displays a wide range of morphologies and size ranges (see Kornmann & Sahling 1980, and references therein).

Distribution: Experimental substrates: 1-30 m stations; background samples: 1 m.

Remarks: The larger cavities encountered as parts of the colonies closely resemble the sporangial cavities as observed by Kornmann & Sahling (1980) in culture. Where several collar-like swellings appear along the main galleries, *Ichnoreticulina elegans* bears similarities to the rhodophyte trace '*Palaeoconchocelis starmachii*'. The genus name *Ichnoreticulina* was most recently introduced by Radtke & Golubic (2005) replacing the established former *Reticulina* due to priority rights (nomen nudum).

Ichnotaxon ***Rhopalia catenata*** Radtke, 1991
Fig. 16A+C
Trace maker ***Phaeophila dendroides*** (Crouan) Batters, 1902
Fig. 16B
Eugomontia sacculata Kornmann, 1960
Fig. 16D

Ichnospecies diagnosis: A tubular system oriented parallel to the substrate surface, often radiating from a central area, with ramifications in pointed angles, with spherical to ellipsoidal swellings along the galleries, bifurcations or gallery terminations, and linked to the substrate surface by thin rhizoidal appendages (after Radtke 1991).

Description: In the present material two morphotypes of *Rhopalia catenata* can be distinguished, the first one (Fig. 16A) being characterised by straight to slightly undulating galleries of varying thickness (3-6 μm) with unilateral or opposite branching in angles of 60-90°. Both, the main gallery and the side branches, exhibit successive irregularly-shaped and flattened swellings, 10-20 μm in diameter. From the latter, characteristic short (20-30 μm) protrusions may emerge. Galleries are often constricted at bifurcations or between individual swellings. In contrast, the galleries of the second morphotype (Fig. 16C) are thicker (7-12 μm), do not exhibit constrictions, and the swellings are less pronounced (ø 10-20 μm). The swellings are not flattened and may occasionally be connected to the substrate surface by short, thin rhizoidal appendages.

Trace maker: The two potential producers of this trace – *Phaeophila dendroides* and *Eugomontia sacculata* – are known for their considerable morphological variability depending on their life cycle (Kornmann 1960), hydrographic parameters (Nielsen 1972; Wilkinson 1974) and host shell conditions (Wilkinson 1975). The only reliable feature to distinguish *Phaeophila dendroides* from *Eugomontia sacculata* is the occasional presence of hairs and its slightly larger diameter. Consequently, the first morphotype can tentatively be assigned to *Eugomontia sacculata* and the second one to *Phaeophila dendroides*.

Distribution: Experimental substrates: 1-30 m stations.

Ichnotaxon ***Eurygonum pennaforme*** Wisshak, Gektidis, Freiwald & Lundälv, 2005
Figs. 16E-G, 32C
Trace maker ***?Epicladia testarum*** (Kylin) Nielsen, 1980
Fig. 16H
Eugomontia sacculata Kornmann, 1960
Fig. 16D

Ichnospecies diagnosis: Colonies oriented closely parallel to the substrate surface, consisting of radiating main galleries from which second- and third-order galleries of uniform diameter bifurcate uni- or bilaterally in angles between 45° and 80° and may exhibit distinct, club-shaped terminal swellings (Wisshak et al. 2005b).

Description: Colonies are oriented close to the substrate surface and exhibit a feather-like overall branching pattern with radiating main galleries from which second- and third-order galleries of uniform diameter (~4-6 μm) and circular outline, uni- or bilaterally bifurcate in angles between 45° and 80°. Branching points may be slightly swollen and the galleries may show distinct, elongated, club-shaped, terminal swellings 6-10 μm in diameter (Fig. 16F-G).

Trace maker: The present material resembles the endolithic part of *Epicladia testarum* as depicted by Nielsen (1987: fig. 5D) from mollusc shells, making this green alga a potential trace maker of *Eurygonum pennaforme*. Whether the morphological variation with the distinct club-shaped terminal swellings is produced by *Epicladia testarum* as well, or by a different *Epicladia* species remains to be answered. *Epicladia testarum* has been reported from the Baltic Sea by Nielsen et al. (1995) among several other species of this genus. Another similar chlorophyte is the vegetative thallus of *Eugomontia sacculata* as depicted by Kornmann (1960: figs. 1A, 6A, F).

Distribution: Experimental substrates: 1-7 m stations; background samples: 1 m.

Remarks: The overall colony shape and branching pattern conforms well to the original diagnosis of *Eurygonum* as given by Schmidt (1992). The trace is clearly distinct from *Eurygonum nodosum*, given the lack of the characteristic nodular appendices, the lower penetration depth and slightly smaller gallery diameters. The trace exhibits some similarity to morphological variations of *Rhopalia catenata*, as reported by Radtke & Golubic (2005: fig. 3E) but is distinguished by the complete absence of rhizoidal appendages, the occasional presence of club-shaped swellings exclusively at the gallery terminations, and otherwise constant gallery diameters. Corresponding traces including the holotype were described under the informal name 'Feder-Form' from the Triassic Cassian Formation by Schmidt (1992) and Glaub & Schmidt (1994), indicating a wide stratigraphic range of the trace. The morphological variation of specimen with and others without the terminal swellings is also clearly expressed in the holotype material. Glaub (1994) depicts a similar Recent trace of *Epicladia testarum* and assigns it to '*Eurygonum* sp.'.

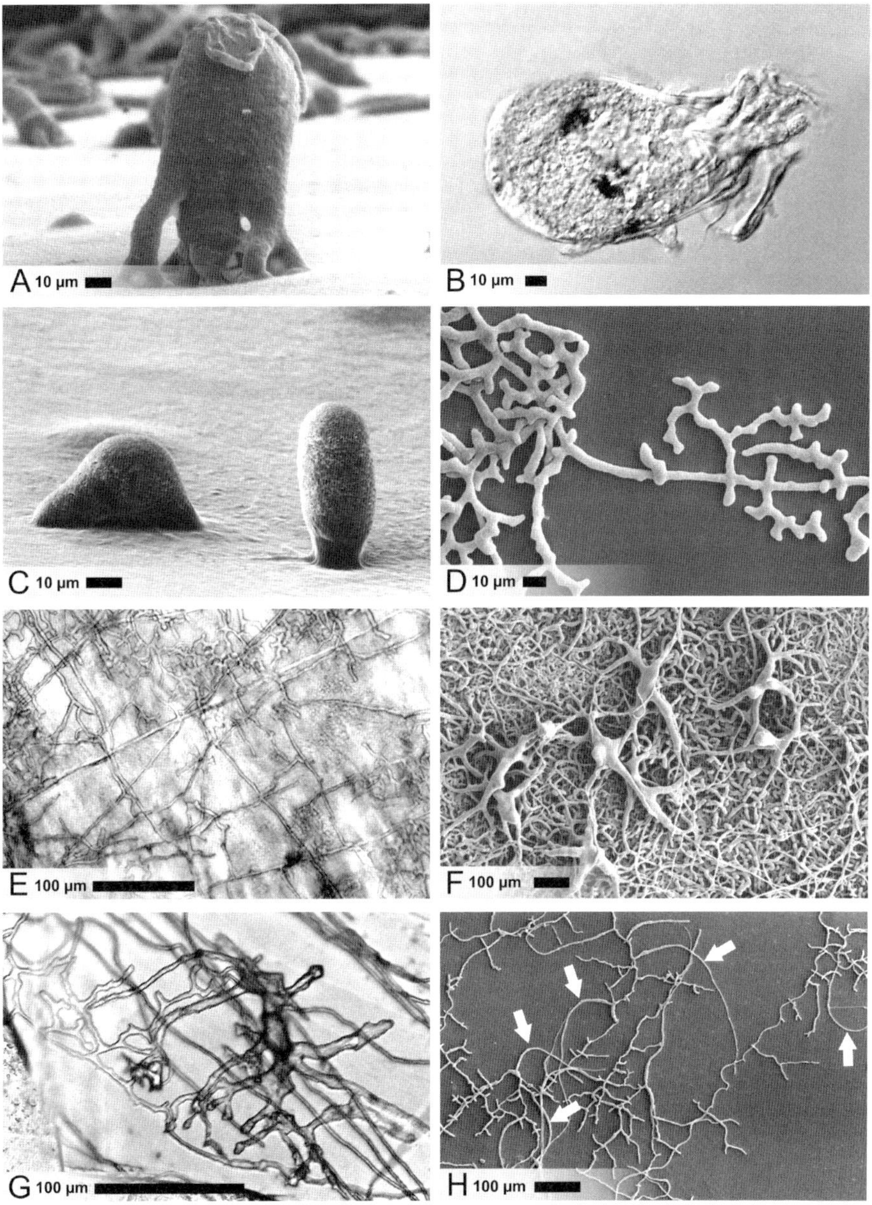

Fig. 15 Traces produced by chlorophytes (SEM) and the corresponding microendoliths (TLM): **A** *Cavernula pediculata* (1 m; 12 months). **B** *Codiolum*-stage of *Gomontia polyrhiza* (1 m L; 6 months; I). **C** Juvenile *C. pediculata* with few or a single rhizoidal appendage (1 m; 6 months). **D** Terminal branching of *Ichnoreticulina elegans* (7 m; 6 months). **E** *Ostreobium quekettii* (6 m; 12 months; BS). **F** Mature network of *I. elegans* with large dendritic chambers (7 m; 15 months). **G** *O. quekettii* with corresponding sporangial cavity (6 m; 12 months; IS). **H** *I. elegans* with whip-shaped galleries (arrows; 7 m; 6 months)

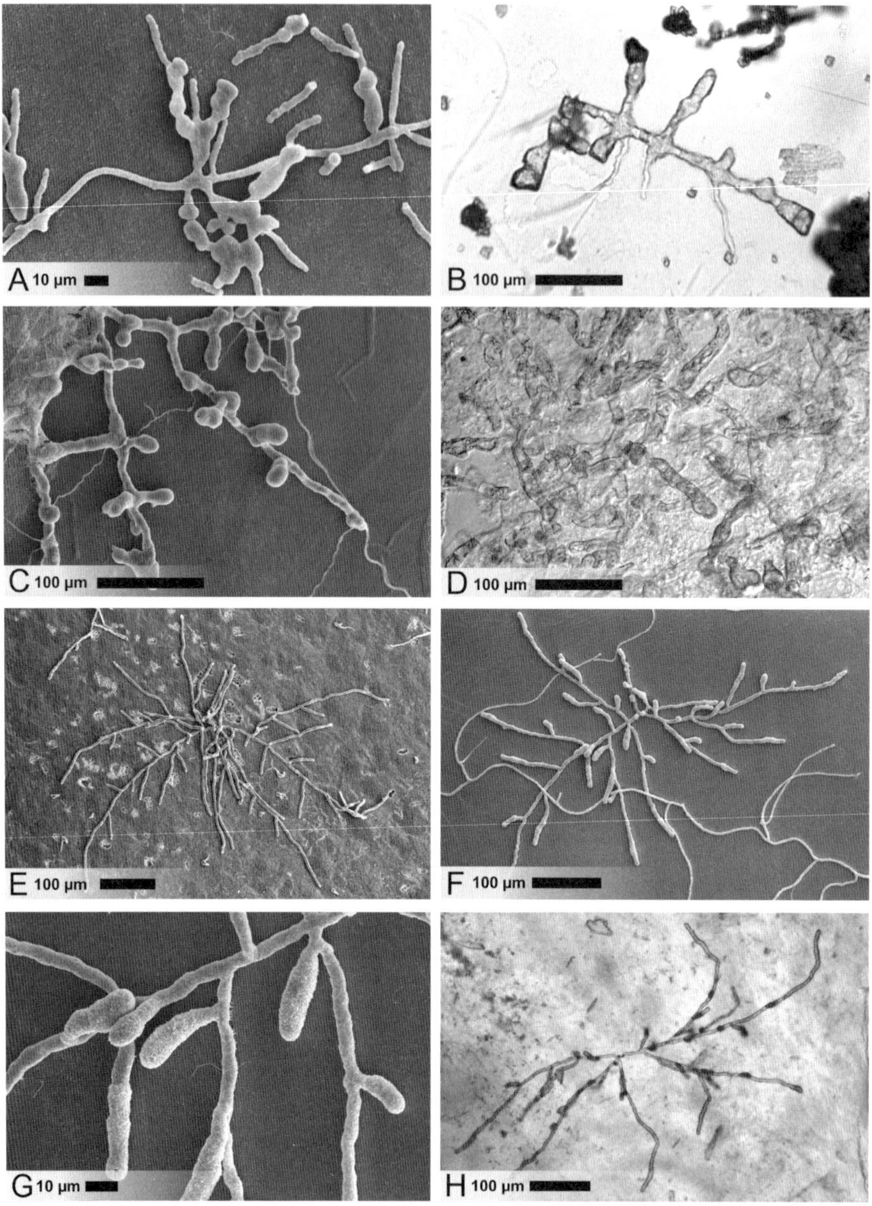

Fig. 16 Traces produced by chlorophytes (SEM) and the corresponding microendoliths (TLM): **A** First morphotype of *Rhopalia catenata* (15 m; 6 months). **B** *Phaeophila dendroides* (1 m L; 6 months; IS). **C** Second morphotype of *R. catenata* (1 m L; 6 months). **D** *Eugomontia sacculata* (1 m L; 18 months; I). **E** *Eurygonum pennaforme* with diagnostic feather-like ramification pattern (7 m; 6 months). **F** Morphotype of *E. pennaforme* with club-shaped terminal swellings (7 m; 6 months). **G** Close-up showing branchings and club-shaped swellings. **H** ?*Epicladia testarum* (1 m L; 6 months; BS)

Fungi

Ichnotaxon ***Saccomorpha clava*** Radtke, 1991
Figs. 19A-B, 27A-E, 32G
Trace maker ***Dodgella priscus*** Zebrowski, 1937

Ichnospecies diagnosis: Club-, sphere- to pear-shaped cavities, interlinked by one or several thin tubes, originating at the base and/or at the main cavity (after Radtke 1991).

Description: Club-shaped sacks (ø 10-30 µm) connected to the surface by narrow (ø 4-6 µm) necks. The sacks are interconnected with evenly thin (ø 2 µm) filaments protruding from the main cavity or at the base of the necks. In the aphotic background material, four different morphotypes could be distinguished: (1) scattered straight (Fig. 27A) or (2) curved (Fig. 27B) individual sacks, 10-20 µm in size, interconnected by few thin galleries, (3) ~30 µm large branched sacks (Fig. 27C) and (4) dense layers of multiple sacks and sack clusters (Fig. 27D) interconnected by a network of galleries.

Trace maker: The ubiquitous fungus *Dodgella priscus* is regarded as the trace maker of this typical trace in aphotic zones by most authors (e.g., Zeff & Perkins 1979; Radtke 1991).

Distribution: Experimental substrates: 30-85 m stations; background samples: 1 and 85 m.

Remarks: It is noteworthy, that a pronounced collar found around the base of the neck in many other occurrences of this trace (e.g., Zeff & Perkins 1979; Beuck & Freiwald 2005) has not been observed in the Kosterfjord material. Early ontogenetic stages of the sacks (ø 4-5 µm; Fig. 19B), which lack interconnecting tunnels, closely resemble the ichnospecies *Cavernula coccidia* and small variants of *Planobola*. Together with the four morphotypes recorded in the background samples from the aphotic zone, these forms possibly represent different ontogenetic stages of the fungal colony.

Ichnotaxon ***Planobola radicatus*** Schmidt, 1992
Fig. 19C
Trace maker ***Dodgella radicatus*** Zebrowski, 1937

Ichnospecies diagnosis: Vertical and spherical cavities, laterally connected to the substrate surface and with proximal radiating galleries (after Schmidt 1992).

Description: Small spherical cavities, approximately 15-20 μm in diameter and penetration depth. From the main cavity, up to 60 μm long, thin (~1 μm) filaments radiate.

Trace maker: The trace resembles the sporangial cavity and radiating filaments of *Dodgella radicatus* as originally described in mollusc shells from Australia by Zebrowski (1937) and later also reported for example from the North Sea (Höhnk 1969) and the North Atlantic (Cavaliere & Alberte 1970).

Distribution: Experimental substrates: 50 m station.

Ichnotaxon ***Saccomorpha terminalis*** Radtke, 1991
Fig. 19D-F
Trace maker ***Phytophthora*** de Bary, 1876
Fig. 19G-H

Ichnospecies diagnosis: Network of fine, branching galleries with thick, oval to spherical terminal cavities (after Radtke 1991).

Description: Two morphotypes of this trace can be distinguished, the first one (Fig. 19D) comprising networks of 2-4 μm thick galleries close to the substrate surface with many characteristic y-shaped, swollen bifurcations. While many branches abruptly end after 10-20 μm, others terminate in 10-100 μm long irregularly sack-shaped cavities. From the latter, thin (~1 μm) filaments may be connected to the substrate surface. The galleries overpass each other frequently within one colony, which reaches an overall extension of several millimetres. The second morphotype (Fig. 19E-F) shows the same characteristic gallery and and y-shaped branching pattern but displays less cavities of more spherical shape. Furthermore, this type is organised in a complex, three dimensional pattern of large (>1 mm) dendritic networks running with some distance parallel to the substrate surface (usually found collapsed on the cast surface). The characteristic y-shaped bi- and trifurcations (Fig. 19F) appear in regular intervals of about 30 μm, with one branch of the trifurcations being connected to the substrate surface.

Trace maker: The trace closely resembles the boring fungus *Phytophthora* sp. as reported from the North Sea in 0-68 m water depth by Höhnk (1969: fig. 4). Corresponding fungi were described by Zebrowski (1937) under the new genus and species name *Arborella kohli*. However, the identity and validity of both taxa bears considerable uncertainties implying the need for a thorough revision (Glaub, pers. comm.).

Distribution: Experimental substrates: 30 and 85 m stations.

Remarks: A similar fossil occurrence was recently reported by Vogel & Marincovich (2004) under the informal name 'Sack-shaped form' (closely resembling the first morphotype) from the Tertiary of Alaska.

Ichnotaxon **'*Flagrichnus*-form 1'**
Figs. 17, 20A-C, 32H
Trace maker ***Schizochytrium*** Goldstein & Belsky, 1964

Description: The traces are usually found collapsed to the cast surface and only partially etched samples reveal their deeply penetrating nature (Fig. 20C). The trace occurs clustered in large numbers of up to several hundred individuals (Fig. 20A). The circular, basal swelling of the traces measures up to 20 μm in diameter and the whip-shaped tube is tapering towards a thin (1-2 μm) filamentous gallery extending straight and deep (up to several 100 μm) into the substrate. Ramifications are rare. Early stages of '*Flagrichnus*-form 1' extend as short thin filament straight into the substrate. The filament progressively grows and thickens as the basal swelling develops (Fig. 17).

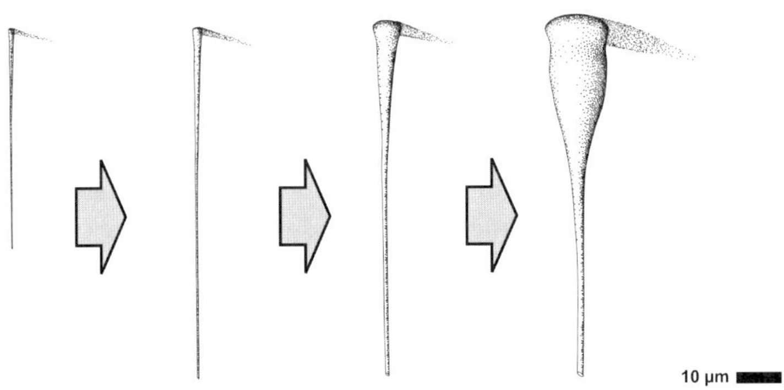

Fig. 17 Schematic sketch illustrating the ontogenetic development of '*Flagrichnus*-form 1' (modified after Wisshak & Porter in press)

Trace maker: Based on histological investigations of corresponding specimens encountered in planted shell fragments in Penobscot Bay, Maine (USA) and Discovery Bay (Jamaica), recovered after six months, Porter & Lingle (1992) and Wisshak & Porter (in press) showed that the trace maker is an eukaryotic zoospore producing heterotroph in the thraustochytrid fungi. Thraustochytrid features are represented by the proximal swollen zoosporangium containing flagellated zoospores, layered cell walls and nuclei with a central nucleolus. The multicellular thallus is pointing towards the genus *Schizochytrium*.

Distribution: Experimental substrates: 7-85 m stations; background samples: 1 m.

Remarks: The formal establishment of the new ichnogenus and ichnospecies and a description of its ontogenetic development were most recently given by Wisshak & Porter (in press). Besides a number of Recent and fossil occurrences reported by the same authors, few occurrences of this microboring are reported in the literature: Two somewhat similar occurrences, but both with more clearly confined basal swellings were reported under the informal names 'Problematic algal form D' (Budd & Perkins 1980: fig. 7D) from shallow waters (0-20 m) at the Puerto Rican shelf, and 'Microboring, Form 5' (Günther 1990: pl. 59, fig. 4-5) from 47 m water depth off Cozumel, Yucatan (Mexico). The to date oldest record of the trace (Glaub pers. comm.) stems from just above the Oligocene/Miocene boundary (Jan Juc Marl / Puebla Clay of the Torquay Basin) from an outcrop on the other side of the globe near Melbourne (Australia). Special attention has to be drawn to the close morphological resemblance with brachiopod punctae in epoxy resin casts. Consequently the identification of '*Flagrichnus*-form 1' in brachiopod shells can only be confirmed with confidence in the case of impunctate brachiopod genera or by the histological identification of the trace maker (in Recent material). The same applies to certain bivalve genera featuring shell tubules, as reported for instance for *Arca*, *Barbatia* and *Glycymeris* by Waller (1980). For *Callista chione* (applied as substrate in the Kosterfjord experiment) tubules are not known and the preexistence of such structures was furthermore excluded by casting and scanning a dozen fresh and complete shells, all of which exhibiting a pristine and smooth inner surface.

Ichnotaxon **'*Flagrichnus*-form 2'**
Figs. 18, 20D-F
Trace maker **unknown chytrid fungus?**

Description: The traces are usually found collapsed to the cast surface and only partially etched shell cross-sections reveal the deeply penetrating nature of these borings (Fig. 20F). The traces occur clustered in large numbers of up to several hundred individuals. At the base of the mature trace, a bilobed to multiply-lobed sack-shaped cavity measuring up to 80 µm in length is developed parallel to the substrate surface. Individual sacks are 10 to 20 µm in diameter and are occasionally connected to the substrate surface by numerous thin filaments (Fig. 20E). From near the base of the sacks, a thin (1-2 µm) filamentous gallery extends straight and deep (up to several 100 µm) into the substrate where it exhibits ramifications and in some cases build up a dense meshwork with those of other individuals. The ontogenetic development of '*Flagrichnus*-form 2' (Fig. 18) starts out as a thin gallery, less than 1 µm in diameter extending straight into the substrate with a pronounced basal swelling about 2 µm in size. From this basal swelling, an initially single but later multiple sack-shaped cavity progressively develops, while the deeply penetrating gallery elongates and ramifies distally. Mature specimens may have sack-shaped cavities up to 80 µm in length with the deeply penetrating part extending more than a millimetre into the substrate. In mature specimen, thin filaments may emerge, connecting the basal cavity to the substrate surface.

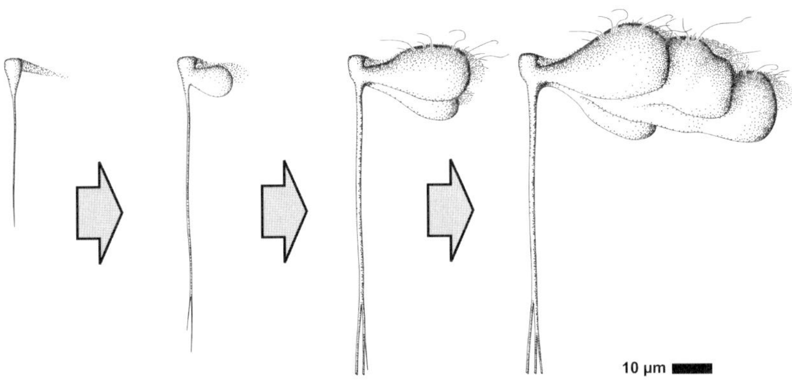

Fig. 18 Schematic sketch illustrating the ontogenetic development of '*Flagrichnus*-form 2' (modified after Wisshak & Porter in press)

Trace maker: Given the bathymetrical distribution of the trace down to aphotic depths, the potential candidates are limited to (chemo)heterotrophic microendoliths and thus either to boring fungi or bacteria. The size and morphology of the trace strongly suggests a fungal origin with the sack-shaped cavities representing sporangia and the deeply penetrating gallery filamentous hyphae or ectoplasmic networks (characteristic of thraustochytrids fungi). Related traces were tentatively interpreted as the work of chytrid fungi by Hook & Golubic (1993: fig. 4) found in bivalves recovered from a deep-sea site at the Florida Escarpment. The formal establishment of this ichnospecies and a discussion on its palaeoenvironmental applicability was recently given by Wisshak & Porter (in press).

Distribution: Experimental substrates: 7-85 m stations.

Remarks: Traces closely resembling the present material were recently reported by Vogel & Marincovich (2004) from the Lower Oligocene Stepovak Formation (Alaska) under the informal name 'Paw-shaped form'. In their description, they note substrate parallel, straight to slightly curved galleries, which represent, most likely, the collapsed deeply penetrating parts of the trace. Juvenile specimens of '*Flagrichnus*-form 2' may be indistinguishable from juvenile '*Flagrichnus*-form 1'. Advanced rosette-like ontogenetic stages of the trace show some affinities to the dendritic ichnospecies *Dendrina belemniticola* from the Cretaceous as depicted by Schnick (1992) which also has a tubular gallery protruding from the centre of the rosette. In the latter case the gallery is not collapsed but originally developed parallel to the substrate surface as revealed by light microscopic analysis (Schnick 1992: pl. IV).

Ichnotaxon ***Orthogonum fusiferum*** Radtke, 1991
Figs. 13A, 20G, 27F, 32F
Trace maker ***Ostracoblabe implexa*** Bornet & Flahault, 1889
Fig. 20H

Ichnospecies diagnosis: Thin, straight to slightly winding, rectangular branching galleries, exhibiting spindle-shaped swellings along the galleries or at branching points (emended after Radtke 1991).

Description: The trace forms a three-dimensional network of thin (~1-2 µm), often rectangular branching galleries, showing characteristic swellings at or between junctions.

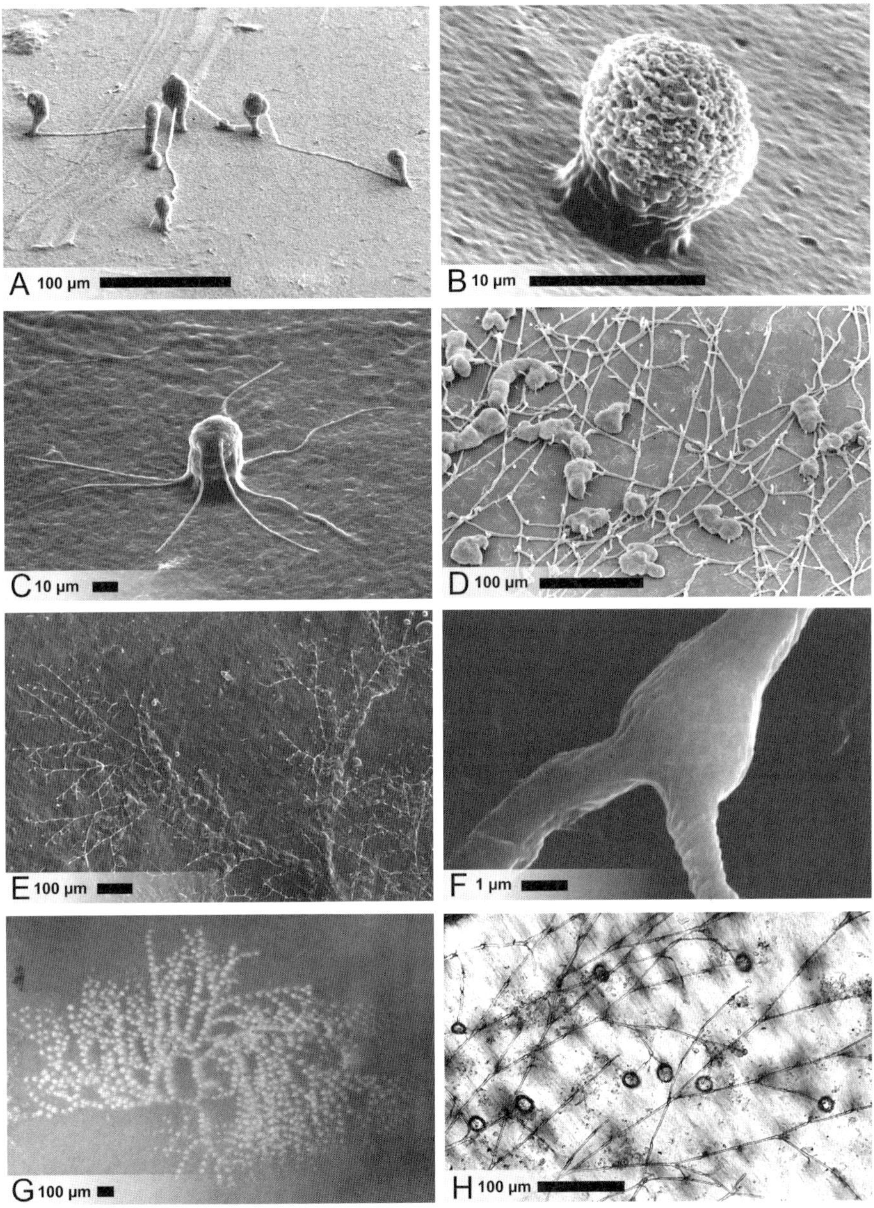

Fig. 19 Traces produced by fungi (SEM) and the corresponding microendoliths (TLM): **A** Small colony of *Saccomorpha clava* (50 m; 12 months). **B** Initial cavity of *S. clava* (30 m; 12 months). **C** *Planobola radicatus* (50 m; 24 months). **D** *Saccomorpha terminalis* with terminal sack-shaped cavities (85 m; 12 months). **E** Part of large *S. terminalis* colony (30 m; 24 months). **F** Detail of diagnostic swellings at ramifications of *S. terminalis*. **G** Large, radiating *Phytophthora* colony (85 m; 6 months; BS).**H** Detail of *Phytophthora* colony with sporangial cavities

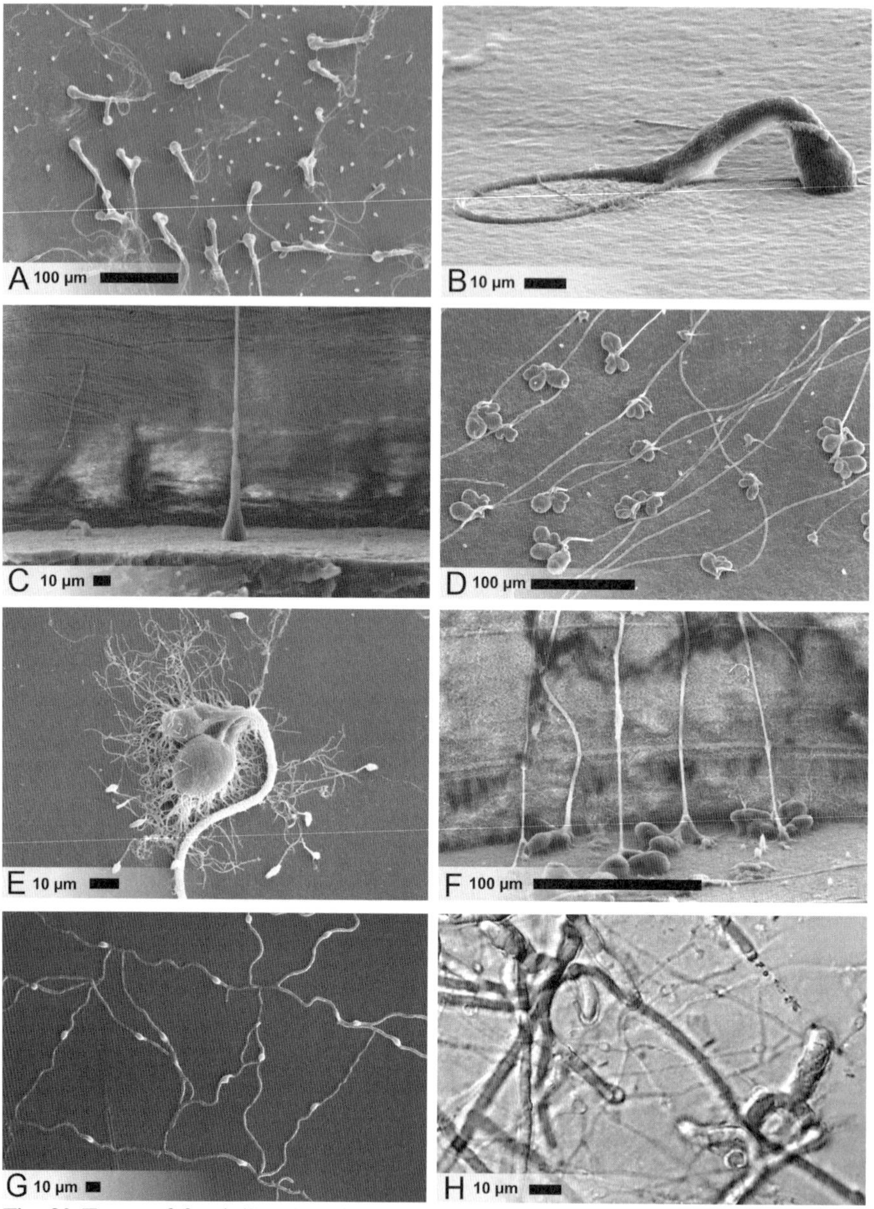

Fig. 20 Traces of fungi (SEM) and the corresponding microendoliths (TLM): **A** Several collapsed ontogenetic stages of '*Flagrichnus*-form 1' (50 m; 6 months). **B** '*Flagrichnus*-form 1' (7 m; 6 months). **C** Partially etched section showing deeply boring habit of '*Flagrichnus*-form 1' (30 m; 24 months). **D** Collapsed '*Flagrichnus*-form 2' (15 m; 6 months). **E** Collapsed '*Flagrichnus*-form 2' (30 m; 12 months). **F** Partially etched section showing deeply boring '*Flagrichnus*-form 2' (15 m; 12 months). **G** *Orthogonum fusiferum* (1 m L; 6 months). **H** *Ostracoblabe implexa* and larger *Mastigocoleus testarum* galleries (1m L; 18 months; I)

Trace maker: The trace is produced by the fungus *Ostracoblabe implexa*, which forms very thin hyphae (1-2 μm) with regular intercalar swellings (2-3 μm). The fungus is of ecological and economical relevance in that it is responsible for epidemic shell diseases of oysters (Alderman & Gareth Jones 1967).

Distribution: Experimental substrates: 1 m station; background samples: 1 and 85 m.

Remarks: The trace is usually found collapsed to the cast surface and then appears as overgrowing other traces or as being oriented closely parallel to the substrate surface. The true deeply penetrating nature of the trace, however, is revealed in partially etched cross sections. Samples from the lagoonal 1 m station, indicate a three dimensional network with a penetration depth of up to many 100 μm. The phrase "parallel to the substrate surface" is consequently omitted from the emended diagnosis here.

Bryozoans

Ichnotaxon ***Pennatichnus*** Mayoral, 1988
Fig. 29A
Trace maker ***Spathipora*** Fischer, 1866

Ichnogenus diagnosis: Ensemble of fine and long tunnels with lateral rounded or drop-shaped primary apertures arranged alternatively, they are connected to the former by short, clearly visible and slightly curved, subordinated conduits of first order. The whole boring system has a very characteristic feather-like appearance; the name is derived from this morphology (after Mayoral 1988).

Description: Elongated cavities (~300-500 μm long at 80-100 μm width) are oriented sub-parallel to the substrate surface and are connected to the surface by a wide aperture (70-90 μm). The cavities may be slightly curved and are proximally connected to a network of evenly thick (10-15 μm) tubular galleries, running closely parallel to the substrate surface.

Trace maker: The distinct traces are produced by the ctenostome bryozoan *Spathipora* with the elongated cavities representing the individual zooids of the colonial organism, connected to a network of stolons by a short pedunculus.

Distribution: Background samples: 85 m.

Remarks: There is a yet unresolved disputation on the formal status as biotaxa versus ichnotaxa of boring bryozoan species and specifically of *Spathipora, Penetrantia* and *Terebripora* (e.g., Boekschoten 1970; Pohowsky 1974, 1978; Mayoral 1988). They were established as biotaxa in the first place but solely based on the morphology of their borings (which in this case can be considered as a perfect mould of the actual organisms) and might thus be regarded as ichnotaxa. Nevertheless, Boekschoten (1970) and Mayoral (1988) offered the alternative ichnotaxa applied here, which are, however, not widely appreciated as yet.

Ichnotaxon ***Iramena*** Boekschoten, 1970
Fig. 29B+F
Trace maker ***Penetrantia*** Silén, 1946

Ichnogenus diagnosis: Borings of probably ctenostome bryozoa, consisting of long (stolon) tunnels in an irregular network, with round to reniform (zooid cavity) apertures situated in alternating positions laterally to and close by the tunnels (Boekschoten 1970).

Description: From a large aperture, individual elongated and tapering cavities extend approximately perpendicular into the substrate with a penetration depth of ~150-250 μm at a width of ~75-125 μm. They are laterally connected to a network of evenly thick (10-15 μm) tubular galleries. This network is oriented in a short distance parallel to the substrate surface, to which it is connected by short perpendicular galleries in regular intervals. The elongated cavities are occasionally associated with distinct spherical cavities ~100 μm in diameter.

Trace maker: The traces are produced by the ctenostome bryozoan *Penetrantia* with the elongated cavities representing the individual zooids of the colonial organism, connected to a network of stolons by a short pedunculus. The spherical cavities are gonozooids.

Distribution: Background samples: 85 m.

Remarks: See *Pennatichnus* above. The traces were misidentified as *Immergentia* in Wisshak et al. (2005a).

Traces of unknown affinity

Ichnotaxon ***Orthogonum lineare*** Glaub, 1994
Figs. 21A-C, 27E-F, 32E
Trace maker **unknown**

Ichnospecies diagnosis: A rectangular branching system of tubes without swellings, with apophyses directed into the substrate, and closely parallel course of the tubes (after Glaub 1994).

Description: Smooth tubes of near-constant diameter (10-15 μm) oriented parallel to the substrate surface. Occasionally, the galleries bear short (>3 μm), spiny protrusions (Fig. 21C). Bifurcations are predominantly rectangular (Fig. 21B). Individual tubes may run parallel to each other.

Trace maker: Although this trace is ubiquitous in aphotic ichnocoenoses, its trace maker is still unknown, but is known to be a heterotroph organism.

Distribution: Experimental substrates: 15-85 m stations; background samples: 1 and 85 m.

Remarks: During early stages of colony development, all specimens consist of long, single and almost straight galleries with only very few rectangular branches. The distinction between *Orthogonum tubulare*, *Orthogonum lineare* and *Orthogonum spinosum*, is somewhat arbitrary and indistinct in the original diagnoses and corresponding specimens even merge laterally into each other quite commonly. Spines as a feature for instance, which are regarded as diagnostic for *Orthogonum spinosum*, are as well exhibited by other ichnotaxa, and thus question the validity of this ichnospecies. The majority of specimens encountered during this study are closest to the diagnosis for *Orthogonum lineare* as given by Glaub (1994) and are distinguished from *Orthogonum tubulare* by more constant tube diameters, the absence of swellings at branchings, and blunt instead of tapering gallery endings.

Ichnotaxon **'*Orthogonum*-form 1'**
Figs. 21D-E, 27G
Trace maker **unknown**

Description: This form comprises thin galleries of near constant diameter (~3-5 μm) that run closely parallel to the substrate surface and describe a wavy course with a wavelength of about 30 μm. Individual tubes can be traced for more than one millimetre. Bifurcations are rare and always rectangular (Fig. 21E). In the background samples from the aphotic zone, the same form was found but with ~10 μm gallery diameter and a respective wavelength of 100-150 μm.

Trace maker: Unknown heterotrophic organism.

Distribution: Experimental substrates: 85 m station; background samples: 85 m.

Remarks: This form corresponds to '*Orthogonum* isp. II' in Wisshak et al. (2005a).

Ichnotaxon **'*Orthogonum*-form 2'**
Fig. 21F
Trace maker **unknown**
Fig. 21G

Description: This form consists of thin (~3 μm) and straight galleries running closely parallel to the substrate surface. Often, several galleries are found to run parallel to each other for many 100 μm but may also cross each other. The galleries only occasionally branch in angles between 45° and 90° and rarely show short lateral swellings.

Trace maker: In a transparent bivalve shell, a heterotrophic endolith was found (Fig. 21G), which probably represents the trace maker of this form. Its taxonomic affinity is not yet determined.

Distribution: Experimental substrates: 15 m station.

Remarks: The trace is found frequently forming dense, substrate-parallel networks in combination with *Ichnoreticulina elegans*. It is distinguished from *Scolecia filosa* by its larger diameter and its orientation closely parallel to the substrate surface. It differs from *Orthogonum lineare* by its smaller gallery diameter and far fewer branching points.

Ichnotaxon **'*Orthogonum*-form 3'**
Fig. 27H
Trace maker **unknown**

Description: Uniformly thick (~10 μm) galleries running not only sub-parallel to the substrate surface but extending typically straight into the substrate, reaching penetration depths exceeding one millimetre. The galleries are smooth or may show spiny protrusions. Branchings are rare and always perpendicular. The deeply penetrating nature of this fungal form becomes obvious in partially etched cross-section casts (Fig. 27H), where the delicate tubes are not collapsed and thus not feigning an orientation parallel to the substrate surface.

Trace maker: Unknown heterotrophic organism.

Distribution: Background samples: 85 m.

Remarks: Further investigations are required to resolve the question whether this form is produced by the same unknown heterotrophic organism that is responsible for *Orthogonum lineare* and may be regarded as a morphological variation of the latter. This form corresponds to '*Orthogonum* isp. I' in Wisshak et al. (2005a).

Ichnotaxon **'Problematic-form 1'**
Fig. 29G
Trace maker **unknown**

Description: Single or fused very thin (<1 μm) fibres, extending straight and deeply into the substrate. Close to the substrate surface, the galleries are ramified and connected to the surface by many minute apertures.

Trace maker: Unknown heterotrophic organism.

Distribution: Background samples: 85 m.

Remarks: The trace is usually found collapsed to the cast surface due to the delicate nature of this trace and their true spatial shape is hard to estimate even in partially etched cross-sectioned casts. In terms of penetration depth and fibrous nature the trace has affinities to *Scolecia filosa*, produced by the cyanobacterium *Plectonema terebrans*, which is, however, characterised by larger diameters, a differing branching pattern and is only encountered in the photic zone.

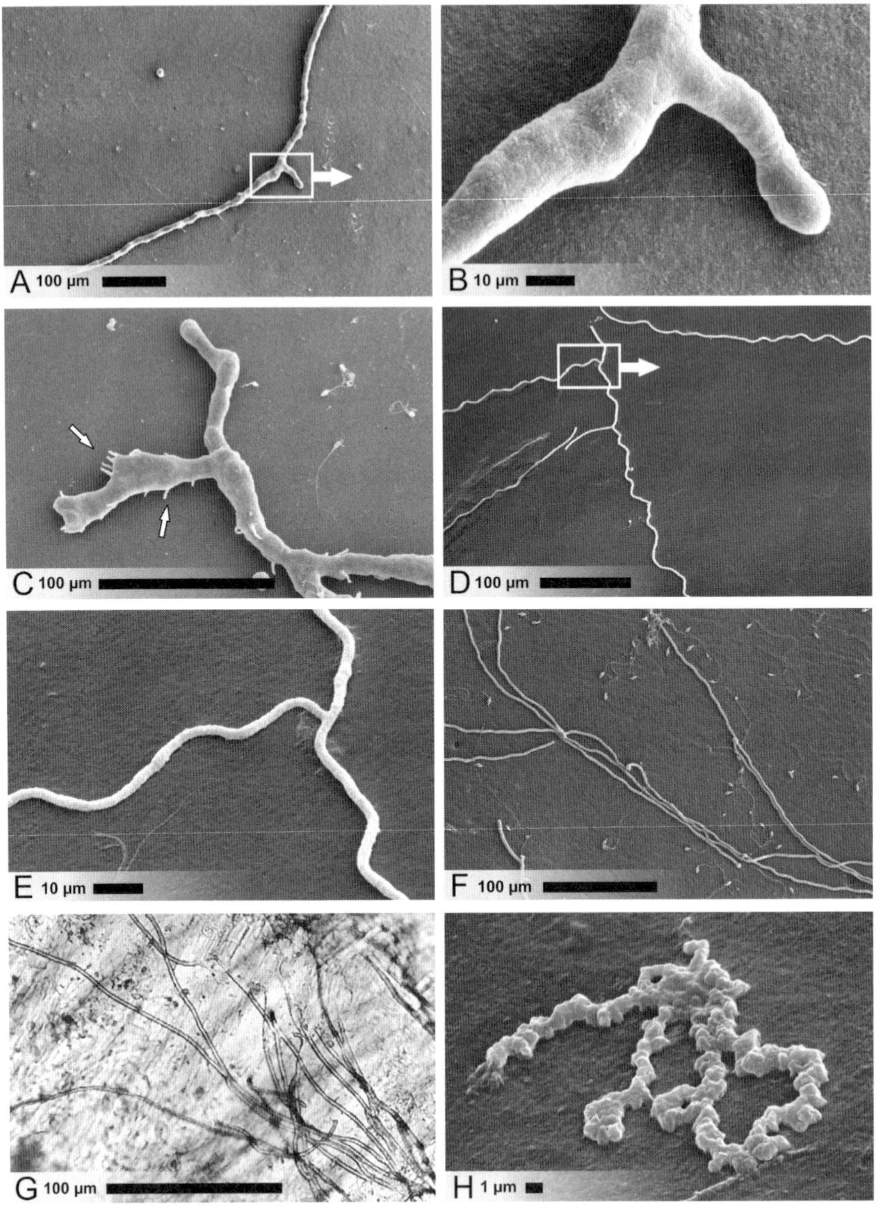

Fig. 21 Traces of unknown affinity (SEM): **A** *Orthogonum lineare* (50 m; 12 months). **B** Close-up of rectangular branching and blunt gallery termination. **C** *Orthogonum lineare* exhibiting short spiny protrusions (arrows) (30 m; 12 months). **D** '*Orthogonum*-form 1' (85 m; 24 months). **E** Close-up of rectangular branching point and uniformly thin wavy gallery. **F** '*Orthogonum*-form 2' (15 m; 12 months). **G** Undetermined trace maker of '*Orthogonum*-form 2' (85 m; 6 months; BS). **H** *Scolecia serrata* (85 m; 24 months)

Ichnotaxon ***Scolecia serrata*** Radtke, 1991
Fig. 21H
Trace maker **unknown**

Ichnospecies diagnosis: Very thin, narrow winding, often serrate and rarely branching galleries close to the substrate surface, running closely besides or on top of each other (after Radtke 1991).

Description: The very thin (~1 μm) galleries with a characteristic serrate micro-sculpture run closely parallel to the substrate surface in narrow windings. The galleries occasionally bifurcate. The trace appears in patchy colonies.

Trace maker: This trace is produced by unknown heterotrophic organisms, most likely filamentous bacteria (Zeff & Perkins 1979; Budd & Perkins 1980).

Distribution: Experimental substrates: 85 m station.

4.2 The macrobioerosion inventory

Traces of macroborers (trace diameter >100 μm) and grazers are of minor importance during early stages of colonisation but were noteworthy diverse and abundant after 2 years exposure (Fig. 22). Most macroboring traces were encountered in the background sample material (see sections 4.3 and 4.4), complementing the following inventory.

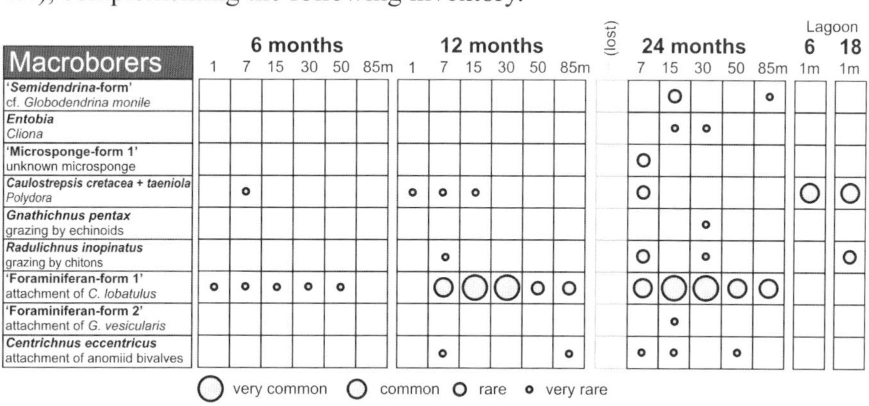

Fig. 22 The occurrence and abundance of macroborers as recorded in epoxy resin casts of *Callista chione* and *Lophelia pertusa* ('*Semidendrina*-form', *Entobia* isp. and 'Microsponge-form 1') and in the micrite bioerosion plates (all others)

On the experimental substrates, shallow etching grooves of foraminiferans were very common at all depths except for the shallowest station with a maximum at 15-30 m of many hundred specimens per dm^2. Two forms were recognised: 'Foraminiferan-form 1' (Fig. 25A-C) and 'Foraminiferan-form 2' (Fig. 25D-E), the former representing chiefly the work of *Cibicides lobatulus* and the latter being produced by the rare *Gypsina vesicularis*. Traces presumably produced by a boring foraminiferan – '*Semidendrina*-form' (Fig. 25F-G) – also occurred, with various ontogenetic stages from juvenile initial borings few tens of micrometre in size up to mature rosettes several 100 µm in diameter. Boring traces of the polychaete *Polydora* resembling the two ichnospecies *Caulostrepsis cretacea* and *Caulostrepsis taeniola* (Fig. 23A-C) were very common at the shallowest stations especially in the lagoonal setting and were rarely found at 7 and 15 m water depth. Grazing traces produced by chitons – *Radulichnus inopinatus* (Fig. 23D) – were found at the shallower stations (1-15 m). Grazing traces produced by echinoids – *Gnathichnus pentax* (Fig. 23E) – were found only at 30 m, where they surround the shallow etching grooves of foraminiferans ('Foraminiferan-form 1') and excavated '*Semidendrina*-form' traces. Attachment scars of anomiid bivalves – *Centrichnus eccentricus* (Fig. 23F) – were rarely found at various depths. In contrast to the background sample material, traces of boring sponges (*Cliona* and other Hadromerida; ichnogenus *Entobia*) were limited to few initial cavities at 15 and 30 m after 2 years exposure (Fig. 23G). Additionally, one yet undescribed type of boring trace ('Microsponge-form 1') was found quite commonly at the 7 m station after 2 years exposure (Fig. 23H). This trace was possibly produced by an unidentified micro-sponge, as indicated by size and micro-sculpture.

In the experimental substrates, 10 ichnotaxa were recorded. In the background samples from aphotic depths, 11 ichnotaxa, and at the shallow-euphotic site 3 ichnotaxa were encountered. In total, 18 different ichnotaxa produced by macroborers or grazers were recorded.

Sponges

Ichnotaxon ***Entobia*** Bronn, 1837
Figs. 23G, 28A-B, 33E-H
Trace maker ***Cliona*** Grant, 1826, and other Hadromerida

Ichnogenus diagnosis: Boring in carbonate substrates comprising a single chamber or networks or boxworks of galleries connected to the surface by several or numerous apertures. Morphology changes markedly with ontogeny. The galleries show progressive increase in diameter during growth; in some forms, inflation at more or less regular distances produces a system of closely interconnected chambers; in other forms, chamber development is restricted to only a brief ontogenetic stage; in still other forms, no cameration is developed. The surface of the boring bears a cuspate microsculpture that may be lost in gerontic specimens. Fine apophyses arise from all or most surfaces of the system (emended diagnosis, Bromley & D'Alessandro 1984).

Description: Entobia is easily recognised in the epoxy resin casts by its comparatively large cavities with a characteristic verrucose surface texture. In the experimental substrates, only initial blackberry-shaped *Entobia* cavities about 100 µm in diameter, connected to the substrate surface by a single aperture (ø ~50 µm), were encountered. In the background sample material a variety of traces were encountered, comprising different ontogenetic stages as well as species variability from solitary small chambers, ~200-300 µm in size (Fig. 33F) to large networks of chambers (Fig. 33G), galleries, apophyses and exploratory threads. The networks are connected to the surface by several apertures. There are principally two different types of micro-sculpture distinguishable: (1) round to oval cells with a rough surface and concentric 'growth lines'. In this type, different stages (resembling the construction of a tiny igloo) record the progressive etching activity by the pseudopodia of the boring cells. (2) Cells with a rather smooth surface showing 3-5 distinct radial ridges and only faint 'growth lines' in between.

Trace maker: Boring sponges, especially relevant of which various species of the genus *Cliona* and few other Hadromerida, are responsible for this most important ichnotaxon in advanced taphonomic stages of shell degradation.

Distribution: Experimental substrates: 15-30 m stations; background samples: 1 and 85 m.

Remarks: For an assignation down to species level, complete mature boring systems (in the case of ichnospecies) or in situ spicules (in the case of sponge species) would be required. A species identification based on surface texture only is not yet feasible.

Ichnotaxon **'Microsponge-form 1'**
Figs. 23H, 28C
Trace maker **unknown poriferan?**

Description: This form is developing irregular cavities, up to several hundred micrometres in size, showing a characteristic verrucose micro-sculpture with individual cells measuring only 3-10 µm. Occasionally, the cells are not fully closed and have a circular opening.

Trace maker: Assigned to an unidentified micro-sponge, based on size and typical verrucose micro-sculpture.

Distribution: Experimental substrates: 7 m station; background samples: 85 m.

Remarks: The trace bears some affinity to the ichnogenus *Entobia*, especially in terms of the verrucose texture, but is much smaller and not multicamerate. Corresponding traces were also found in skeletons of *Desmophyllum dianthus* from the Reñihue Fjord, southern Chile (Försterra et al. 2005: fig. 9C).

Ichnotaxon **'Microsponge-form 2'**
Fig. 28D
Trace maker **unknown poriferan?**

Description: 'Microsponge-form 2' shows a central cavity, few tens to 100 µm in diameter, from which radial thin (~5 µm), bifurcating tunnels emerge sub-parallel to the substrate surface.

Trace maker: Tentatively assigned to an unidentified micro-sponge.

Distribution: Background samples: 85 m.

Remarks: This form resembles 'Sponge, Form 3' reported by Günther (1990) from a Recent reef setting in Yucatan (Mexico).

Ichnotaxon **'Microsponge-form 3'**
Fig. 28E
Trace maker **unknown poriferan?**

Description: 'Microsponge-form 3' are large sack-shaped traces with diameters of 100-200 µm and penetration depths of 150-200 µm, that are connected to the substrate surface by one or few thick rhizoidal appendages.

Trace maker: Tentatively assigned to an unidentified micro-sponge.

Distribution: Background samples: 85 m.

Remarks: This form shows some features diagnostic for the ichnospecies *Cavernula pediculata*, but can be distinguished from the latter by its relatively large size.

Ichnotaxon **'Microsponge-form 4'**
Fig. 28F
Trace maker **unknown poriferan?**

Description: This form is highly variable in its morphological appearance and develops irregularly-shaped, branching to dendritic borings, 100-1,000 μm in size, with characteristic whip-shaped appendages up to 100 μm in length.

Trace maker: Tentatively assigned to an unidentified micro-sponge.

Distribution: Background samples: 85 m.

Remarks: This form shows close similarities to 'Entobia-Form 3' (Glaub 1994), 'Sponge, Form 1' (Günther 1990), 'Echinoid form' (Radtke 1993), 'Sponge form B' (Budd & Perkins 1980) and possibly 'Spinate boring form' (Zeff & Perkins 1979). The bristle-like processes also show an affinity to '*Semidendrina*-form' (see below).

Ichnotaxon **'Microsponge-form 5'**
Fig. 28G
Trace maker **unknown poriferan?**

Description: The trace exhibits a morphological variability, ranging from 40-60 μm large isolated or linked spherical aggregates to dendritic irregularly-shaped boring systems close to the substrate surface that reach more than 1 mm in size. They have a verrucose to granular micro-sculpture.

Trace maker: Tentatively assigned to an unidentified micro-sponge, based on its size, multicamerate architecture and surface sculpture.

Distribution: Background samples: 85 m.

Remarks: Budd & Perkins (1980: fig. 9F) depict similar aggregates, 40-50 μm in size, but tentatively interpret them as fungal sporangia.

Ichnotaxon '**Microsponge-form 6**'
Fig. 28H
Trace maker **unknown poriferan?**

Description: 'Microsponge-form 6' is a shallow dendritic boring system, in places more than 1 mm in diameter, characterised by a central, irregularly-shaped flat area, from which many ramifying branches radiate and rapidly decrease in width.

Trace maker: Tentatively assigned to an unidentified micro-sponge.

Distribution: Background samples: 85 m.

Remarks: The form bears some similarities to the fossil ichnogenus *Platydendrina* as described by Vogel et al. (1987) from Devonian strata.

Polychaetes

Ichnotaxon ***Caulostrepsis taeniola*** Clarke, 1908
Figs. 23A-C, 33A-C
Trace maker ***Polydora*** Bosc, 1802

Ichnospecies diagnosis: Gallery cylindrical, bent in a narrow U which is sometimes enlarged in the shape of a tongue. The inward-facing margins of the limbs are always interconnected by a distinct vane. Limbs closer or partially fused towards the apertural extremity. Transverse section dumbbell-shaped, aperture 8-shaped (emended diagnosis, Bromley & D'Alessandro 1983).

Description: Different ontogenetic stages of the borings are recorded, ranging from small initial stages to patches with numerous and partially stacked specimens. The traces do not necessarily lie within one plane, but may curve transversally or perpendicular to the length axis. Intersections were not observed. Individual traces comprise a cylindrical gallery bent in an u-shaped tongue from the aperture(s), with the inward-facing margins of the gallery being interconnected by a vane. The traces measure up to more than one centimetre in length and the width of the cylindrical galleries ranges from 150-200 μm in initial borings to 500 μm in mature ones. The surface texture of the trace may reflect the substrate ultrastructure, taking on a foreign sculpture.

Trace maker: The spionid polychaete genus *Polydora* is the best-documented producer of Recent *Caulostrepsis*. In the Kosterfjord area, *Polydora ciliata* is the chief representative of this genus.

Distribution: Experimental substrates: 1-15 m stations; background samples: 1 m.

Remarks: Both, *Caulostrepsis taeniola* and *Caulostrepsis cretacea* (see below) are encountered. They are primarily distinguished by the presence or absence of a thin vane. Intermediate morphological variants, however, are common and the vane is more clearly developed in the proximal region of the traces, rendering a clear distinction difficult or impossible. The apertural region may not always be 8-shaped, but may diverge into two separate galleries bending outward and entering the substrate sub-parallel. This feature is morphologically related to *Maeandropolydora decipiens* as described by Voigt (1965). Where specimens are densely spaced, a morphological affinity to *Caulostrepsis contorta* is apparent.

Ichnotaxon ***Caulostrepsis cretacea*** (Voigt, 1971)
Fig. 33D
Trace maker ***Polydora*** Bosc, 1802

Ichnospecies diagnosis: Galleries bent in a long, narrow U-form with the inward-facing walls of the limbs fused by complete removal; the original position of the median wall is sometimes indicated by a very shallow axial depression along the structure. Vane absent. Transverse section always flattened-elliptical but showing gradual decrease in width toward the aperture. Shape of aperture flattened-oval (emended diagnosis, Bromley & D'Alessandro 1983).

Description: The morphology of this species closely resembles that of *Caulostrepsis taeniola* (see above), except for the vane being almost or as thick as the main tunnel.

Trace maker: See *Caulostrepsis taeniola*.

Distribution: Experimental substrates: 1-15 m stations; background samples: 1 m.

Remarks: This trace is morphologically related to, and often difficult to tell apart from *Caulostrepsis taeniola* (see above).

Echinoids

Ichnotaxon ***Gnathichnus pentax*** Bromley, 1975
Fig. 23D
Trace maker **regular echinoids**

Ichnospecies diagnosis: Gnathichnus consists of a regular stellate grouping of five similar grooves radiating at c. 72° (Bromley 1975).

Description: The pentaradiate patches of grooves of this trace measure usually less than a millimetre in diameter. Individual narrow and straight grooves measure 50-200 µm in length.

Trace maker: This distinct trace is produced by the teeth of regular echinoids, browsing on boring and encrusting organisms on hard substrates. In the present case, the preferred association with foraminiferal traces suggest a somewhat selective preying behaviour of the involved echinoids. The comparatively small size of the present *Gnathichnus* of less than 1 mm is inferring rather small or juvenile echinoids. The generally more typical appearance of this trace as pavements of overlapping stars (Bromley 1975) was not encountered on the experimental substrates.

Distribution: Experimental substrates: 30 m station.

Bivalves

Ichnotaxon ***Centrichnus eccentricus*** Bromley & Martinell, 1991
Fig. 23E
Trace maker **anomiid bivalves**

Ichnospecies diagnosis: Shallow biogenic etching traces on carbonate lithic or skeletal substrates comprising centrically arranged arcuate or ring-shaped grooves (Bromley & Martinell 1991).

Description: The shallow etching scars measure 1 to 2 mm in length and comprise a pear-shaped, off-centre array of centrically-curved grooves.

Trace maker: The trace was found to be produced as shallow attachment scars by the byssus of anomiid bivalves (Bromley & Martinell 1991). In the Kosterfjord, small anomiid bivalves (chiefly *Heteranomia squamula*) are common and were often found attached to the experimental frames and substrates.

Distribution: Experimental substrates: 7-15 m and 50-85 m stations.

Chitons

Ichnotaxon ***Radulichnus inopinatus*** Voigt, 1977
Fig. 23F
Trace maker **chitons**

Ichnospecies diagnosis: Meandering paths consisting of densely packed longitudinal patches or grooves side by side with straight parallel fine furrows or scratches engraved into the substratum (Voigt 1977).

Description: The trace is characterised by dense patches, up to several cm^2 in area, comprising small clusters of parallel to sub-parallel minute scratches. Individual scratches are up to 200 μm in length and often slightly meandering.

Trace maker: As reflected in the name *Radulichnus*, this trace is known to be produced by the rasping action of herbivorous gastropods and/or chitons. The present material closely resembles the Recent material figured in the original description by Voigt (1977), who found chitons (e.g., *Lepidochiton*) to be the most likely trace makers. Traces produced by limpets in contrast, show more regular clusters of parallel straight scratches as reported for instance for *Acmaea* by Akpan (1984), and the clusters produced by *Littorina* follow a distinct meandering arrangement (Ankel 1936, 1937), that was not observed in the present material.

Distribution: Experimental substrates: 1-7 m and 30 m stations.

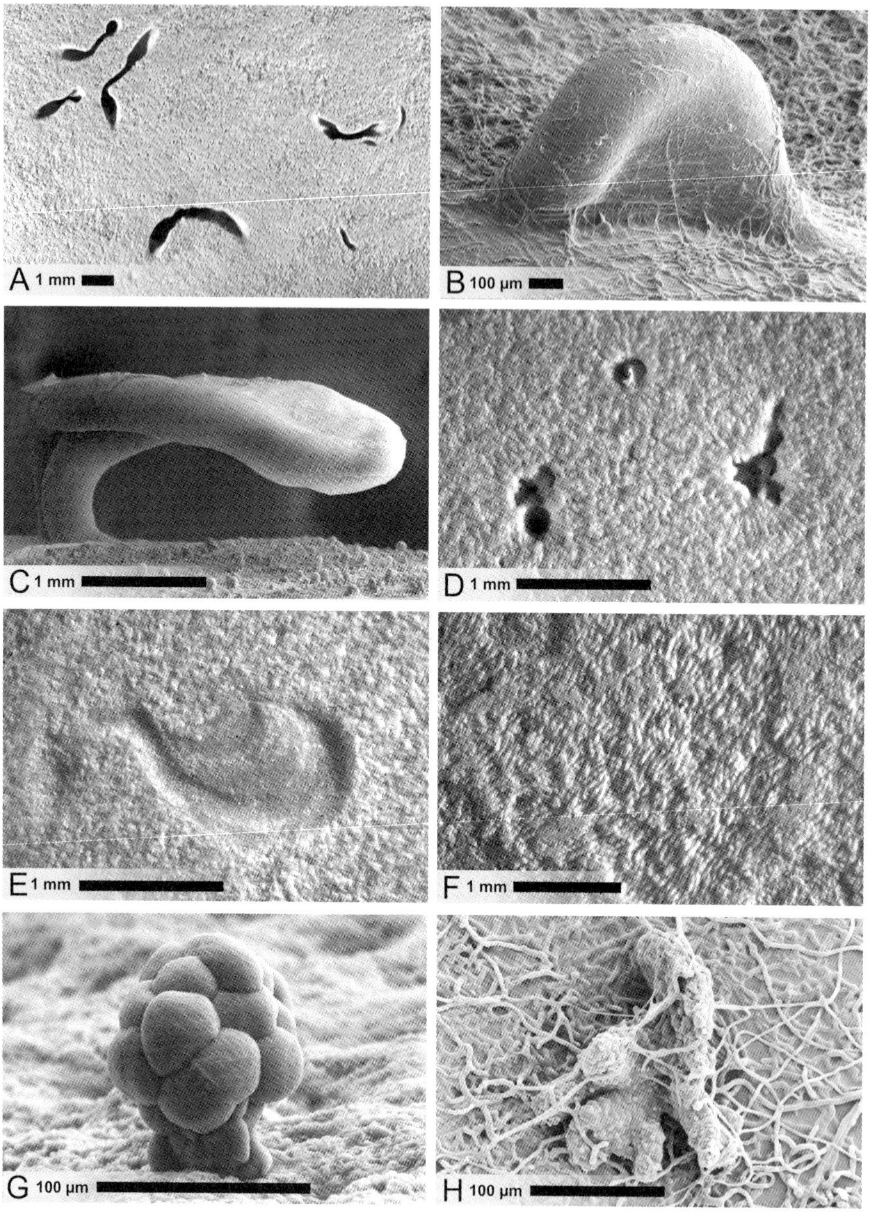

Fig. 23 Traces produced by macroborers (SEM; **A**, **D**-**F**: ILM): **A** Carbonate plate with apertures of polychaete borings *Caulostrepsis* (1 m; 24 months). **B** Initial *Caulostrepsis* (1 m L; 24 months). **C** Mature *C. cretacea* (1 m L; 6 months). **D** Gastropod grazing trace *Radulichnus inopinatus* (7 m; 24 months). **E** Echinoid gnawing trace *Gnathichnus pentax* excavating '*Semidendrina*-form' (7 m; 24 months). **F** Anomiid attachment scar *Centrichnus eccentricus* (50 m; 24 months). **G** Sponge boring *Entobia* (85 m; 24 months). **H** 'Microsponge-form 1' (7 m; 12 months)

Foraminiferans

Ichnotaxon **'*Semidendrina*-form'**
Figs. 24, 25F-H, 29C-D
Trace maker **cf. *Globodendrina monile*** Plewes, Palmer & Haynes, 1993

Description: Large (~0.5-1 mm), complex boring system, comprising an off-centre main chamber connected to the substrate surface by a single aperture (30-60 µm wide), and a plexus of branching and anastomosing irregular galleries issuing from one side of the main chamber and running substrate-parallel. The galleries may show long, the main cavity short whip-shaped apophyses (ø max. 2 µm) and/or a verrucose surface texture. In mature specimen, the plexus may describe a half circle around the main chamber.

Trace maker: The foraminiferal origin of '*Semidendrina*-form' is currently under disputation. Corresponding borings were interpreted as the work of an unknown endolithic foraminiferan by Cherchi & Schroeder (1991) who found foraminiferal tests in the main chamber of some specimen on radiographs. In a different approach, the producing foraminiferan was named *Globodendrina monile* by Plewes et al. (1993) based on Jurassic fossil material with small agglutinated chimneys around the entrance of the borings. The latter feature was interpreted as an evolutionary reduced remnant of an agglutinating foraminiferan and the new species was consequently placed in the order Astorhizida. Only recently, the trace itself was ichnotaxonomically treated and the foraminiferal origin was discussed and questioned after neither the endolithic tests nor agglutinating chimneys were found in an extensive Recent material (Bromley et al. in press). However, they found some evidence still pointing towards a foraminiferan as producer of the trace (e.g., the anastomosing branching pattern) and discuss the possibility of a naked type of foraminiferan as likely trace maker. Based on the Kosterfjord material, Bromley et al. (in press) were able to show an ontogenetic series, comprising small initial main chambers with short lateral protrusions to large specimen with a mature plexus (Fig. 24).

Distribution: Experimental substrates: 15 and 85 m stations; background samples: 85 m.

Remarks: The formal establishment of a corresponding new ichnogenus and ichnospecies, as well as a detailed discussion of the biological interpretation and the geologic record of this form and related traces is given in Bromley et al. (in press).

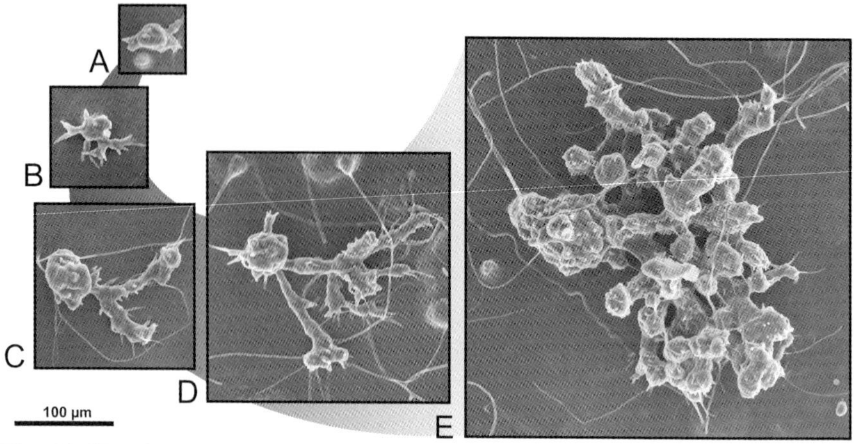

Fig. 24 SEM images of an ontogenetic series of '*Semidendrina*-form' as recorded in an epoxy resin cast of a single square centimetre of a planted *Callista* shell, that was exposed for a two year period in 15 m water depth (modified after Bromley et al. in press). **A-E** Five ontogenetic stages showing progressive development of the fan-shaped plexus

Ichnotaxon **'Foraminiferan-form 1'**
Fig. 25A-C
Trace maker e.g., ***Cibicides lobatulus*** (Walker & Jakob, 1798)
Fig. 40A-B

Description: The circular to sub-circular shallow depressions measure up to more than 1 mm in diameter and often clearly express a spiral pattern with the central whirls showing deepest relief. Other variants are smaller, more circular in outline and lack a spiral pattern. In epoxy resin casts of the traces a rough surface texture becomes visible, contrasting the smooth ambient surface of the shell (Fig. 25B) and amplifying pre-existing boring structures of microendoliths (Fig. 25C).

Trace maker: The very abundant shallow attachment scars are primarily produced by the ubiquitous foraminiferan *Cibicides lobatulus*. The smaller and more circular variants may as well be produced by other rotaliinaid foraminiferans.

Distribution: Experimental substrates: 1-85 m stations.

Remarks: Even though resembling attachment scars were repeatedly reported in the literature (e.g., Vénec-Peyré 1996; Bromley 2005), a formal ichnotaxonomic treatment is yet to be undertaken.

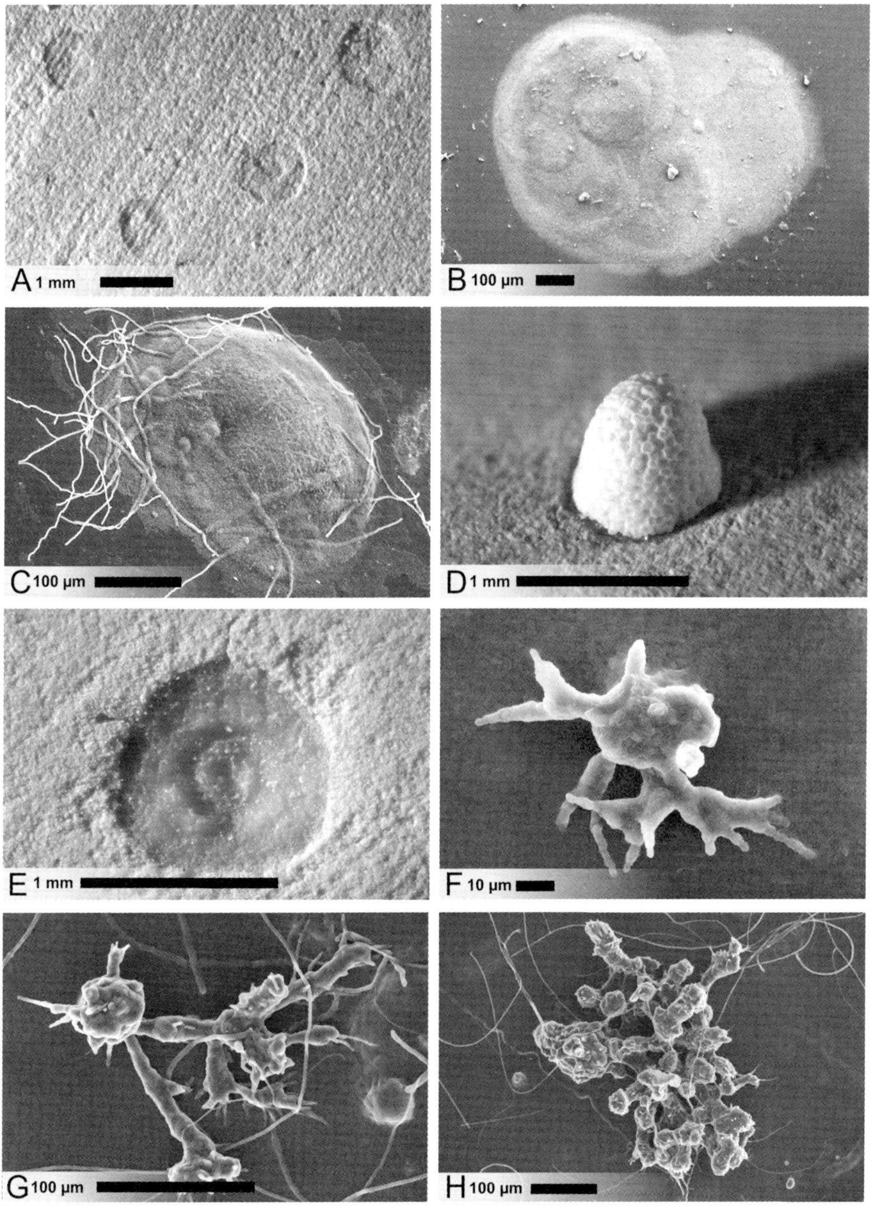

Fig. 25 Traces produced by foraminiferans (SEM; **A**, **D-E**: ILM): **A** *Cibicides lobatulus* etching scars 'Foraminiferan-form 1' reflecting spiral growth (15 m; 24 months). **B** 'Foraminiferan-form 1' with rough etching texture contrasting the smooth ambient surface (85 m; 12 months). **C** 'Foraminiferan-form 1' amplifying pre-existing microborings (15 m; 12 months). **D** *Gypsina vesicularis* etching 'Foraminiferan-form 2' (15 m; 24 months). **E** 'Foraminiferan-form 2' with central depression and short radiating grooves (15 m; 24 months). **F-H** Ontogenetic series of '*Semidendrina*-form' (15 m; 24 months)

Ichnotaxon **'Foraminiferan-form 2'**
Fig. 25D-E
Trace maker ***Gypsina vesicularis*** (Parker & Jones, 1860)
Figs. 25D, 40L-M

Description: Circular etching scars, 0.8-1 mm in diameter, with a characteristic central depression and occasionally some short radiating grooves.

Trace maker: This rare attachment trace is produced by the rotaliid foraminiferan *Gypsina vesicularis*, which was observed in situ in its trace on the experimental carbonate plates (Fig. 25D).

Distribution: Experimental substrates: 15 m station.

Ichnotaxon **'Foraminiferan-form 3'**
Fig. 29E
Trace maker ***Hyrrokkin sarcophaga*** Cedhagen, 1994

Description: This form is composed of irregular clusters (ø up to 1.5 mm) of several dozen tapering galleries up to 250 μm in length each, radiating deeply into the substrate. The surface texture of the occasionally bifurcating tunnels is typically found to be rough and may show a weakly developed incrementation.

Trace maker: The trace was originally interpreted as brachiopod attachment scar in Wisshak et al. (2005a). However, more recent investigations by Beuck and López Correa (pers. comm.) clearly identify the trace as etching of the parasitic rosalinid foraminiferan *Hyrrokkin sarcophaga*.

Distribution: Background samples: 85 m.

Remarks: Hyrrokkin sarcophaga or their traces have repeatedly been reported from fossil and Recent *Lophelia* skeletons (Freiwald & Schönfeld 1996; Beuck & Freiwald 2005; Bromley 2005).

Brachiopods

Ichnotaxon ***Podichnus centrifugalis*** Bromley & Surlyk, 1973
Fig. 29F
Trace maker ***Macandrevia cranium*** (Müller, 1776)

Ichnospecies diagnosis: More or less compact group of pits or cylindrical holes in hard, calcareous substrates. The pits at the centre of the group more or less perpendicular to the surface, the more periferal pits typically deeper and larger, entering the substrate obliquely, centrifugally. Size of pits up to ca. 200 µm (Bromley & Surlyk 1973).

Description: The traces are characterised by a cluster (ø ~350 µm) of around 20 tapering intrusions. The central ones are shorter and approximately perpendicular to the surface while the more periferal ones are longer and enter the substrate obliquely and centrifugally.

Trace maker: Podichnus is known to be produced as etched attachment scars left by the pedicle of brachiopods (Bromley & Surlyk 1973). The abundant brachiopod *Macandrevia cranium* was recorded in the same sample material and is most likely responsible for the traces encountered.

Distribution: Background samples: 85 m.

Traces of unknown affinity

Ichnotaxon **'Problematic-form 2'**
Fig. 29H
Trace maker **unknown**

Description: 'Problematic-form 2' is a relatively large trace (up to 600 µm in size), comprising clusters of several straight and tapering intrusions (20-150 µm), penetrating the substrate perpendicularly from an irregularly-shaped shallow depression (elevation in the cast).

Trace maker: The trace possibly represents an attachment scar of a heterotroph epibiont, such as a brachiopod or a foraminiferan.

Distribution: Background samples: 85 m.

Remarks: Corresponding traces were found in skeletons of *Desmophyllum dianthus* from the Reñihue Fjord, southern Chile (Försterra et al. 2005: fig. 6G-H).

4.3 Bioerosion at the *Lophelia* reef site

In order to get insight into the ichnocoenoses beared by advanced stages of shell degradation, skeletons of *Lophelia pertusa* and few associated bivalves collected at the Säcken Reef site were analysed for endolithic traces. The cold-water coral *Lophelia* was chosen because of being a sessile organism (preventing bathymetrical bias) and for enabling a comparison with other *Lophelia* bioerosion studies (Krutschinna 1997; Beuck & Freiwald 2005). In total, 19 different traces produced by fungi (5), boring sponges (7), bryozoans (2), foraminiferans (2) and brachiopods (1) were found. In addition, two traces of unknown affinities are recorded (Table 7).

All traces encountered in this aphotic setting are produced by heterotrophic organisms, most prominent of which are boring fungi. Important and ubiquitous ichnospecies are *Saccomorpha clava* (Fig. 27A-E) produced by the fungus *Dodgella priscus*, and *Orthogonum lineare* (Fig. 27E-F), whose trace maker is still unknown but is expected to be found among the boring fungi as well. Both traces are the key constituents of the *Saccomorpha clava / Orthogonum lineare* ichnocoenosis (Fig. 27E) which is well-developed in most *Lophelia* samples. All further fungal or potential fungal traces were only rarely encountered as there are *Orthogonum fusiferum* (Fig. 27F; trace maker *Ostracoblabe implexa*), '*Orthogonum*-form 1' (Fig. 27G; unknown producer) and '*Orthogonum*-form 2' (Fig. 27H; unknown producer).

Among the sponge borings, only *Entobia* (Fig. 28A-B; produced predominantly by clionaid sponges and other Hadromerida) is a dominating ichnotaxon. The boring systems of *Entobia* are especially ubiquitous in advanced taphonomic stages, where they may remove more than two thirds of the skeletal material (Freiwald & Wilson 1998). Abandoned sponge cavities are often subject to subsequent infestation of various microendoliths. Besides *Entobia*, a number of rare potential micro-sponge borings ('Microsponge forms 1-6'; Fig. 28C-H) were encountered alongside traces only tentatively assignable to this group based on their large size and/or micro-sculpture.

Boring bryozoans are represented by two ichnogenera – the very common *Pennatichnus* (Fig. 29A) produced by the ctenostome *Spathipora*, and the far less abundant *Iramena* (Fig. 29B) standing for the biotaxon *Penetrantia*.

Foraminiferal traces are restricted to very few specimen of '*Semidendrina*-form' (Fig. 29C-D), possibly produced by an agglutinating or naked foraminiferan. In addition, the attachment scars 'Foraminiferan-form 3' of the parasitic rosalinid *Hyrrokkin sarcophaga* were recorded (Fig. 29E).

Attachment scars left by the pedicle of brachiopods – represented by the ichnospecies *Podichnus centrifugalis* (Fig. 29F) – were most likely produced by *Macandrevia cranium*, which is very common in the Kosterfjord area.

Table 7 List of ichnotaxa and their known or assumed producers as well as their relative abundance as recorded in epoxy resin casts of *Lophelia pertusa* and associated bivalves recovered from the Säcken Reef site in 85 m water depth (+ + very common, + common, - rare, - - very rare)

Ichnotaxa	Trace maker	Abund.
Saccomorpha clava	*Dodgella priscus* (fungus)	+ +
Orthogonum lineare	? (unknown heterotroph, ?fungus)	+ +
Orthogonum fusiferum	*Ostracoblabe implexa* (fungus)	-
'*Orthogonum*-form 1'	? (unknown heterotroph)	-
'*Orthogonum*-form 3'	? (unknown heterotroph)	+
Entobia	*Cliona* and other Hadromerida (sponge)	+ +
'Microsponge-form 1'	? (micro-sponge)	- -
'Microsponge-form 2'	? (micro-sponge)	- -
'Microsponge-form 3'	? (micro-sponge)	- -
'Microsponge-form 4'	? (micro-sponge)	- -
'Microsponge-form 5'	? (micro-sponge)	- -
'Microsponge-form 6'	? (micro-sponge)	- -
Pennatichnus	*Spathipora* (bryozoan)	+ +
Iramena	*Penetrantia* (bryozoan)	-
'*Semidendrina*-form'	cf. *Globodendrina monile* (foraminiferan)	- -
'Foraminiferan-form 3'	*Hyrrokkin sarcophaga* (foraminiferan)	- -
Podichnus centrifugalis	*Macandrevia cranium* (brachiopod)	- -
'Problematic-form 1'	? (unknown heterotroph)	-
'Problematic-form 2'	? (unknown heterotroph)	-

Last but not least, two enigmatic forms were recognised, the first one of which ('Problematic-form 1'; Fig. 29G) is similar to *Scolecia filosa* but is extending deeply into the substrate with a smaller gallery diameter and ramifications near the substrate surface. Their spatial shape is hard to determine even in partially etched cross-sections due to minute gallery diameters. The other form ('Problematic-form 2'; Fig. 29H) is possibly an attachment scar of a heterotroph epibiont, such as a brachiopod or a foraminiferan.

When examining thin-sections of dead *Lophelia* skeletons, the nature of the micritic envelope as a dense layer of microborings becomes evident (Fig. 26), contrasting the deeply penetrating large sponge borings (Fig. 26B). A close-up of this darker layer reveals the individual borings, most prominent of which the galleries of the unknown producer of *Orthogonum lineare* (Fig. 26C) and the sporangial cavities of the fungus *Dodgella priscus* (Fig. 26D). In addition, these images clearly show the limits of light microscopy, underlining the superior suitability of the epoxy resin cast / SEM methodology for revealing the three-dimensional architecture of microendolithic traces.

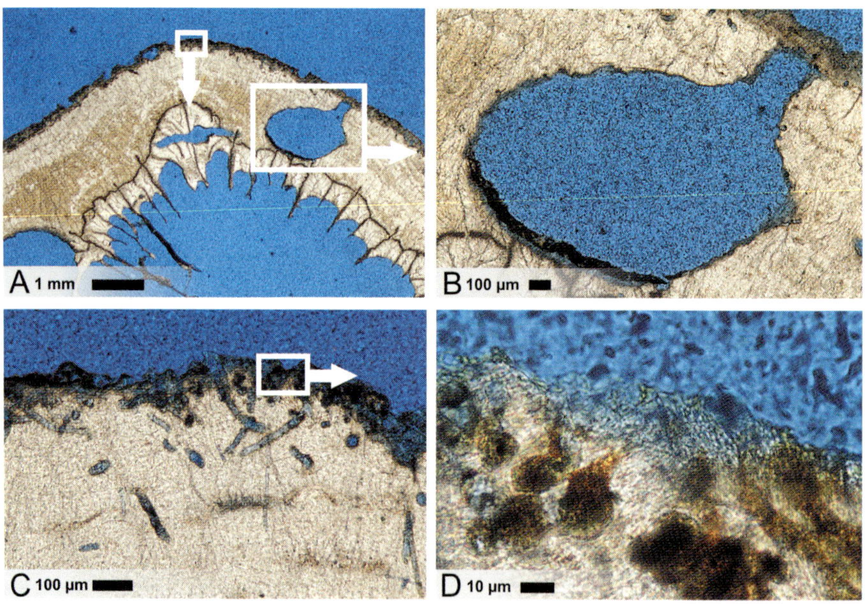

Fig. 26 Thin-section of a dead *Lophelia pertusa* skeleton from the Säcken Reef site. **A** Overview of a *Lophelia* cross-section (outside upward facing, septae downward facing) with a distinct micritic envelope and large cavities of boring sponges. **B** Close-up of sponge boring with large apertural canal. **C** Close-up of the micritic envelope consisting of a dense layer of microborings. **D** Close-up of sporangial cavities belonging to the fungus *Dodgella priscus*

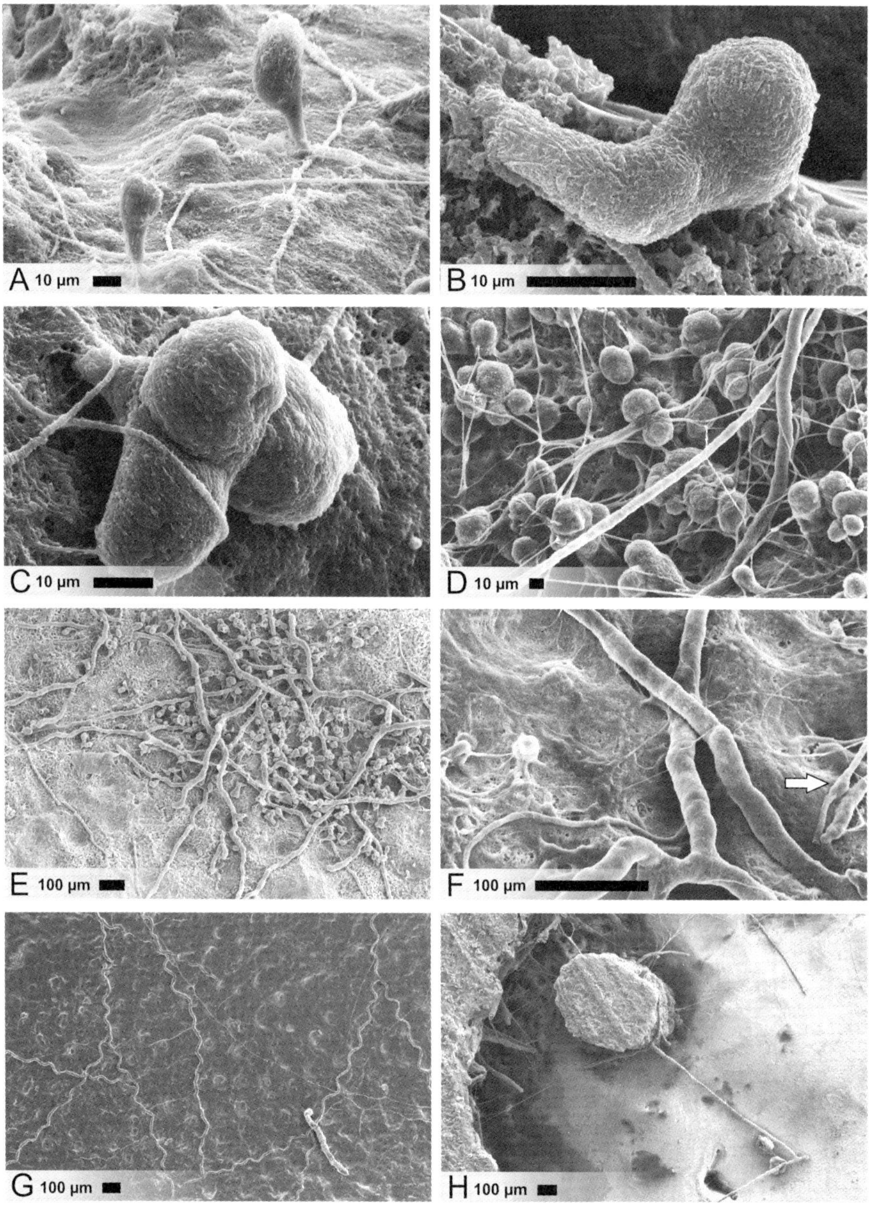

Fig. 27 Traces produced by fungi (SEM) recorded in *Lophelia* skeletons from the Säcken Reef site: **A-D** Different morphotypes of *Saccomorpha clava* **A** Sporangia-bearing cavities connected by thin tubular galleries. **B** Curved variation. **C** Branched sacks. **D** Dense clusters of sacks. **E** *Orthogonum lineare* in typical association with *S. clava*. **F** Thick tubes of *O. lineare* and thin galleries of *Orthogonum fusiferum* with characteristic swellings (arrow). **G** Diagnostic wavy galleries of '*Orthogonum*-form 1'. **H** Deeply penetrating '*Orthogonum*-form 3' in a partially etched section

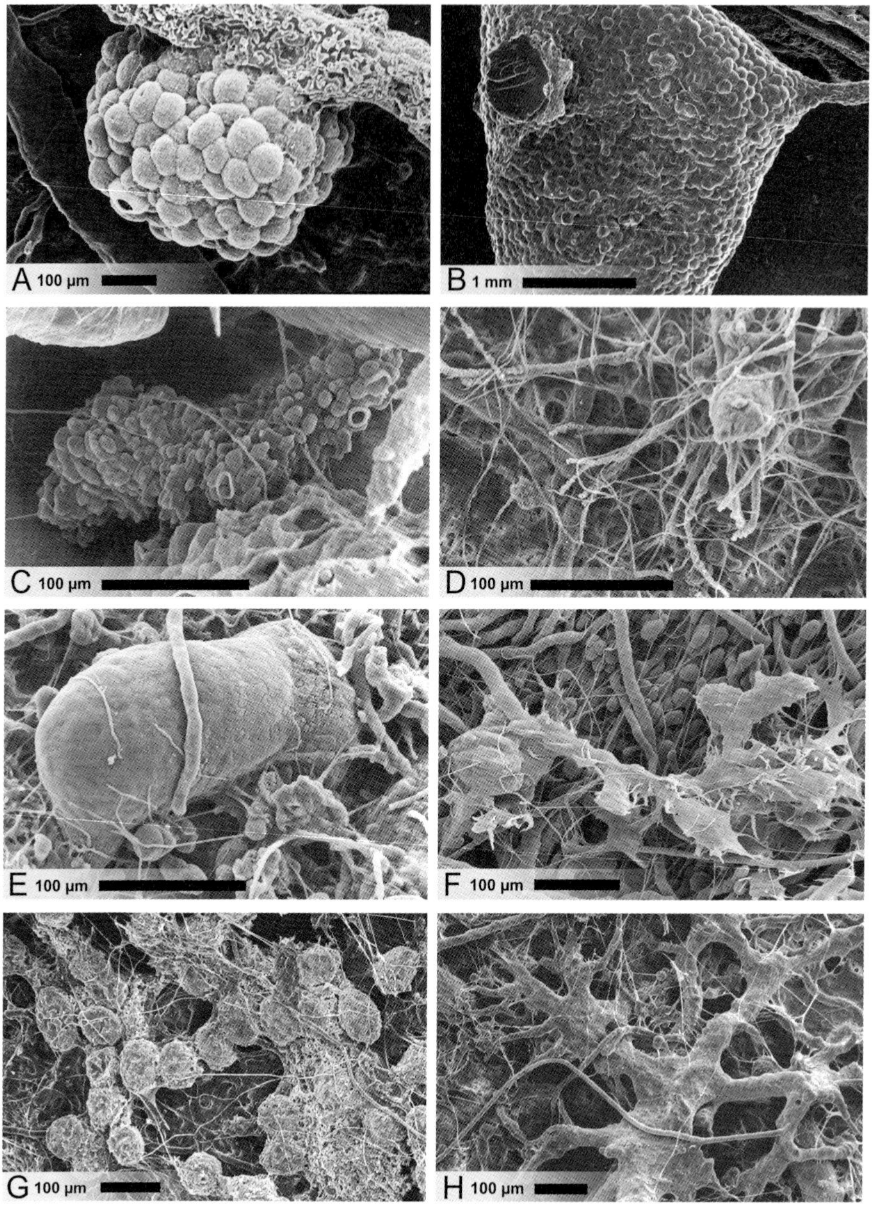

Fig. 28 Traces produced by boring sponges (SEM) recorded in *Lophelia* skeletons sampled at the Säcken Reef site: **A** Small *Entobia* with verrucose microsculpture and a boring cell 'caught in action' (lower left). **B** Large *Entobia* chamber. **C** 'Microsponge-form 1' with tiny, irregularly-shaped boring cells. **D** 'Microsponge-form 2' with bifurcating exploratory threads. **E** 'Microsponge-form 3' with thick rhizoidal appendages. **F** Irregularly-shaped 'Microsponge-form 4' with bristle-like protrusions. **G** Spherical to irregular flat aggregates of 'Microsponge-form 5'. **H** Dendritic 'Microsponge-form 6'

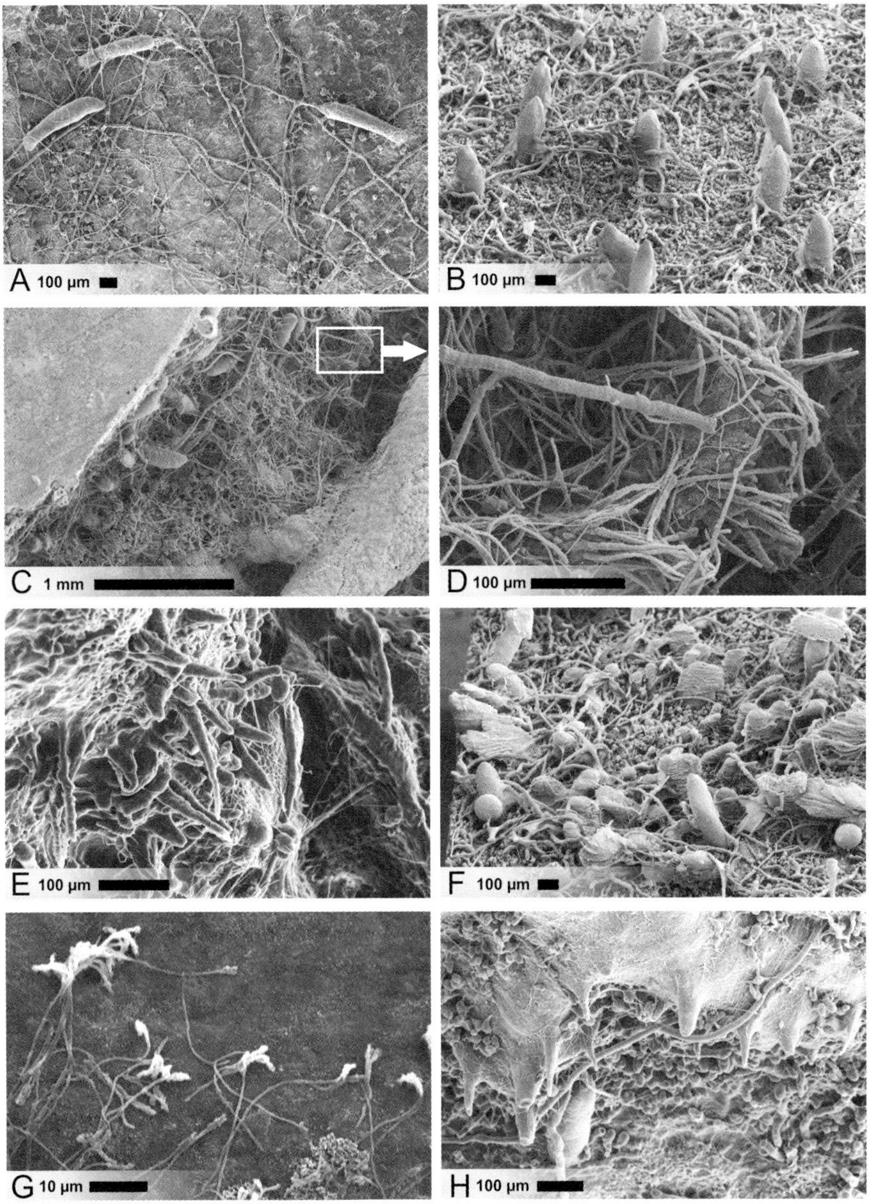

Fig. 29 Traces produced by bryozoans (**A**-**B**), foraminiferans (**C**-**E**), brachiopods (**F**) and unknown trace makers (**G**-**H**) recorded in *Lophelia* skeletons from the Säcken Reef site (SEM): **A** *Iramena* with cavities interconnected by thin galleries. **B** *Pennatichnus* with perpendicular cavities. **C** Partly hidden ?foraminiferal trace '*Semidendrina*-form'. **D** Close-up of whip-shaped protrusions. **E** *Hyrrokkin* etching 'Foraminiferan-form 3'. **F** Attachment scar *Podichnus centrifugalis.* **G** Collapsed deeply penetrating 'Problematic-form 1'. **H** 'Problematic-form 2' with intrusions originating from a shallow depression

4.4 Bioerosion of intertidal *Littorina* shells

While the Säcken Reef samples record the ichnocoenoses in aphotic depths, the shallow-water snail *Littorina littorea* was chosen as a bathymetrically restricted representative of the shallow-euphotic end member. Several dozen *Littorina* shells in various stages of degradation were collected at the deployment site of the shallowest experimental panels and were analysed for endolithic traces. In total, 17 different traces produced by cyanobacteria (7), chlorophytes (3), fungi (3), boring sponges (1) and polychaetes (2) were found and one trace of unknown affinity is recorded (Table 8).

Already on a first macroscopic sight, the important contribution of bioerosion agents in the breakdown of *Littorina* shells becomes evident in form of the distinct u-shaped borings (*Caulostrepsis taeniola* and *Caulostrepsis cretacea*), primarily produced by various *Polydora* species (Fig. 30A-B). Most *Littorina* shells show traces of infestation by this polychaete, and advanced taphonomic stages bear densely-spaced borings that are often unroofed by mechanical abrasion. These traces are exclusively found intruding from the shells exterior and never penetrate to the interior, suggesting a syn-vivo infestation. Also clearly recognisable on a macroscopic scale but less abundant are the multicamerate boring systems (*Entobia*) mainly produced by clionaid sponges (Fig. 30C). They are recognised by evenly spread apertures leading to a sometimes also unroofed subsurface network of interlinked chambers and exploratory threads.

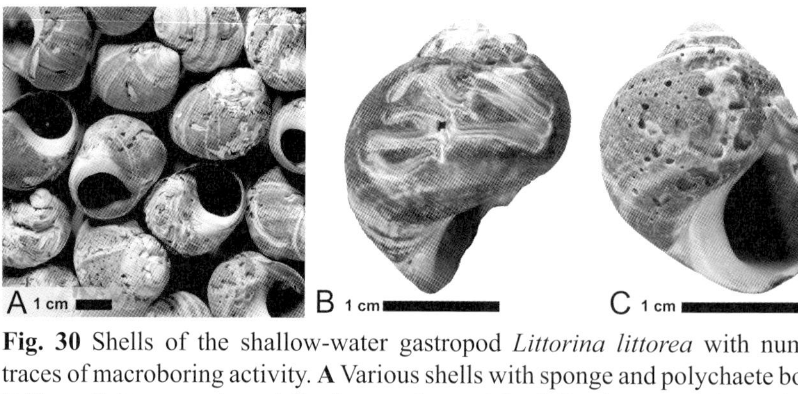

Fig. 30 Shells of the shallow-water gastropod *Littorina littorea* with numerous traces of macroboring activity. **A** Various shells with sponge and polychaete borings. **B** Unroofed tongue-shaped *Caulostrepsis taeniola*. **C** Partly unroofed *Entobia*

On a microscopic scale, the characteristic verrucose micro-sculpture of *Entobia* (e.g., Fig. 33H) and the details of colony architecture (Fig. 33G) can be studied. In one case, the very initial boring attempt of a sponge was recorded in form of a polygonal pattern of incomplete boring cells, penetrating the substrate for only few micrometres (Fig. 33E).

Table 8 List of ichnotaxa and their known or assumed producers as well as their relative abundance as recorded in epoxy resin casts of *Littorina littorea* collected from the shoreline near the Säcken Reef site (+ + very common, + common, - rare, - - very rare)

Ichnotaxa	Trace maker	Abund.
Fascichnus dactylus	e.g., *Hyella caespitosa* (cyanobacterium)	+ +
Fascichnus frutex	*Hyella gigas* (cyanobacterium)	+
Fascichnus rogus	cf. *Hyella racemus* (cyanobacterium)	-
'*Fascichnus*-form 1'	*Hyella caespitosa* (cyanobacterium)	-
Eurygonum nodosum	*Mastigocoleus testarum* (cyanobacterium)	-
Planobola	cf. *Cyanosaccus piriformis* (cyanobacterium)	+
Scolecia filosa	*Plectonema terebrans* (cyanobacterium)	+ +
Cavernula pediculata	*Gomontia polyrhiza* (chlorophyte)	+ +
Eurygonum pennaforme	?*Epicladia testarum* (chlorophyte)	+
Ichnoreticulina elegans	*Ostreobium quekettii* (chlorophyte)	-
Orthogonum lineare	? (fungus)	- -
Orthogonum fusiferum	*Ostracoblabe implexa* (fungus)	+
Saccomorpha clava	*Dodgella priscus* (fungus)	- -
'*Flagrichnus*-form 1'	*Schizochytrium* (fungus)	-
Caulostrepsis taeniola	*Polydora* (polychaete)	+ +
Caulostrepsis cretacea	*Polydora* (polychaete)	+ +
Entobia	*Cliona* and other Hadromerida (sponge)	+

Various ontogenetic stages (Fig. 33A-C) of the ubiquitous *Caulostrepsis* ichnospecies are encountered. In terms of microboring activity, the ichnocoenosis is largely dominated by cyanobacterial traces, most dominant of which being *Fascichnus dactylus* (Fig. 31A; trace maker for instance *Hyella caespitosa*) and the filamentous *Scolecia filosa* (Fig. 31H; produced by *Plectonema terebrans*). Less abundant but still commonly found are *Fascichnus frutex* (Fig. 31B; produced by *Hyella gigas*) and *Planobola* (Fig. 31G; probably representing the work of unicellular cyanobacteria). Further rare cyanobacterial traces are represented by *Eurygonum nodosum* (Fig. 31F; trace maker *Mastigocoleus testarum*) and the '*Fascichnus*-form 1' (Fig. 31E; produced by *Hyella caespitosa*). Both traces were found to be more abundant towards slightly deeper waters (7 m station) on the experimental substrates. A finding that was not recorded in the experimental substrates was that of the distinct raspberry-shaped aggregates of *Fascichnus rogus* (Fig. 31C-D), a trace originally only reported from presumable shallow-marine tropical carbonates of Silurian to Jurassic age (Bundschuh & Balog 2000). The present material is extending the stratigraphical range of this ichnospecies to the Recent and is questioning the assumed producer *Hyella racemus*, which is only known from Recent tropical settings.

Among the boring traces of chlorophytes, only *Cavernula pediculata* (Fig. 32A-B; trace maker *Gomontia polyrhiza*) reaches a dominant abundance and exhibits a variable morphology ranging from narrow deeply penetrating to more shallow but wide variants. Also quite commonly found are the feather-like branching colonies of *Eurygonum pennaforme* (Fig. 32C; produced possibly by *Epicladia testarum* or *Eugomontia sacculata*). *Ichnoreticulina elegans* (Fig. 32D; trace maker *Ostreobium quekettii*) appears to be rare in the *Littorina* shells.

Boring fungi play only a comparatively minor role as microbioerosion agents and are predominantly represented by the delicate colonies of *Orthogonum fusiferum* (Fig. 32F; trace maker *Ostracoblabe implexa*) and '*Flagrichnus*-form 1' (Fig. 32H; trace maker *Schizochytrium*). Both traces are deeply penetrating forms, found collapsed to the cast surface apparently overgrowing other traces. *Saccomorpha clava* (Fig. 32G; trace maker *Dodgella priscus*) is very rare in this shallow-water substrate, as is the assumed fungal boring *Orthogonum lineare* (Fig. 32E).

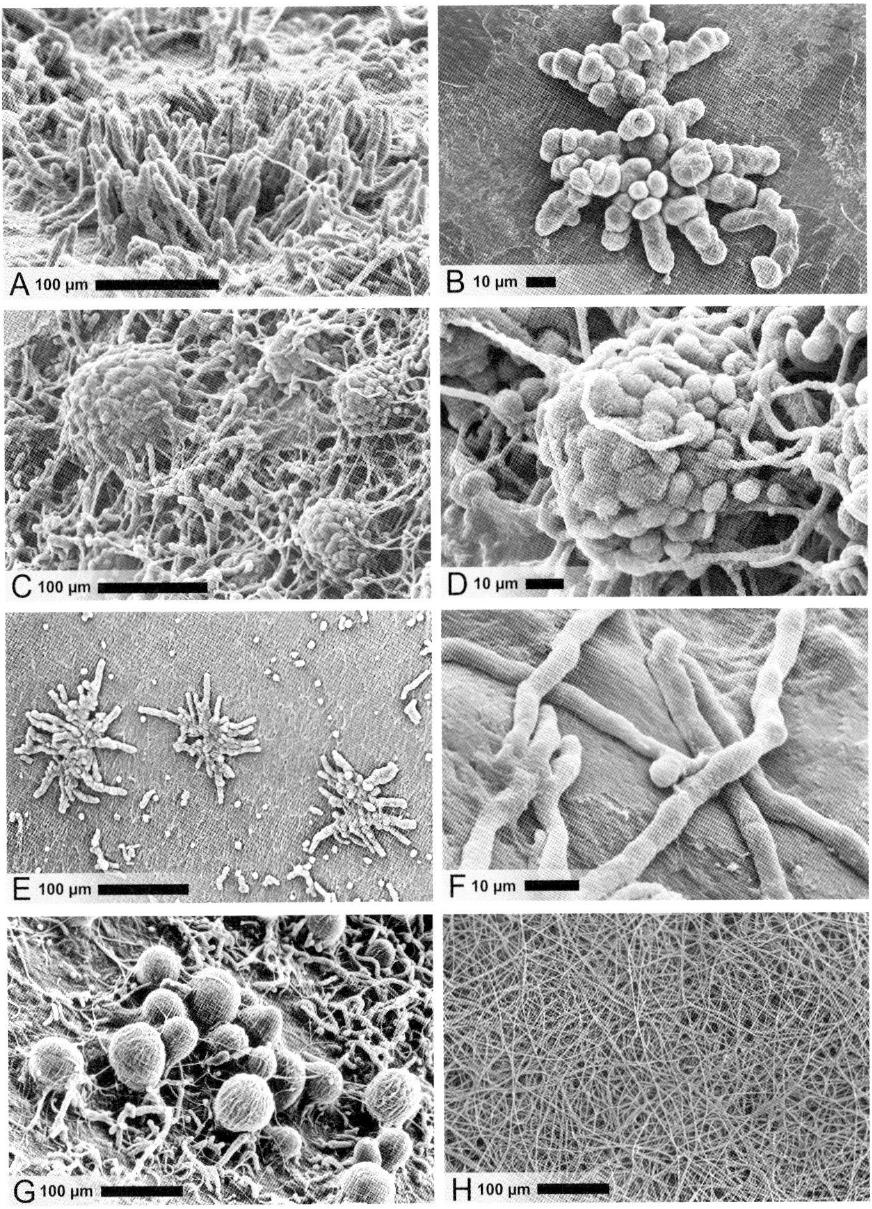

Fig. 31 Traces produced by cyanobacteria (SEM) recorded in *Littorina* shells collected at the shoreline: **A** Typical bundle of thin galleries characterising the trace *Fascichnus dactylus*. **B** Small *Fascichnus frutex* boring with cell-like gallery structure. **C-D** Overview and close-up of raspberry-shaped *Fascichnus rogus* aggregates. **E** Three '*Fascichnus*-form 1' colonies running closely parallel to the substrate surface. **F** Few galleries of *Eurygonum nodosum*. **G** The spherical aggregates of *Planobola*. **H** Partly collapsed dense carpet of *Scolecia filosa*

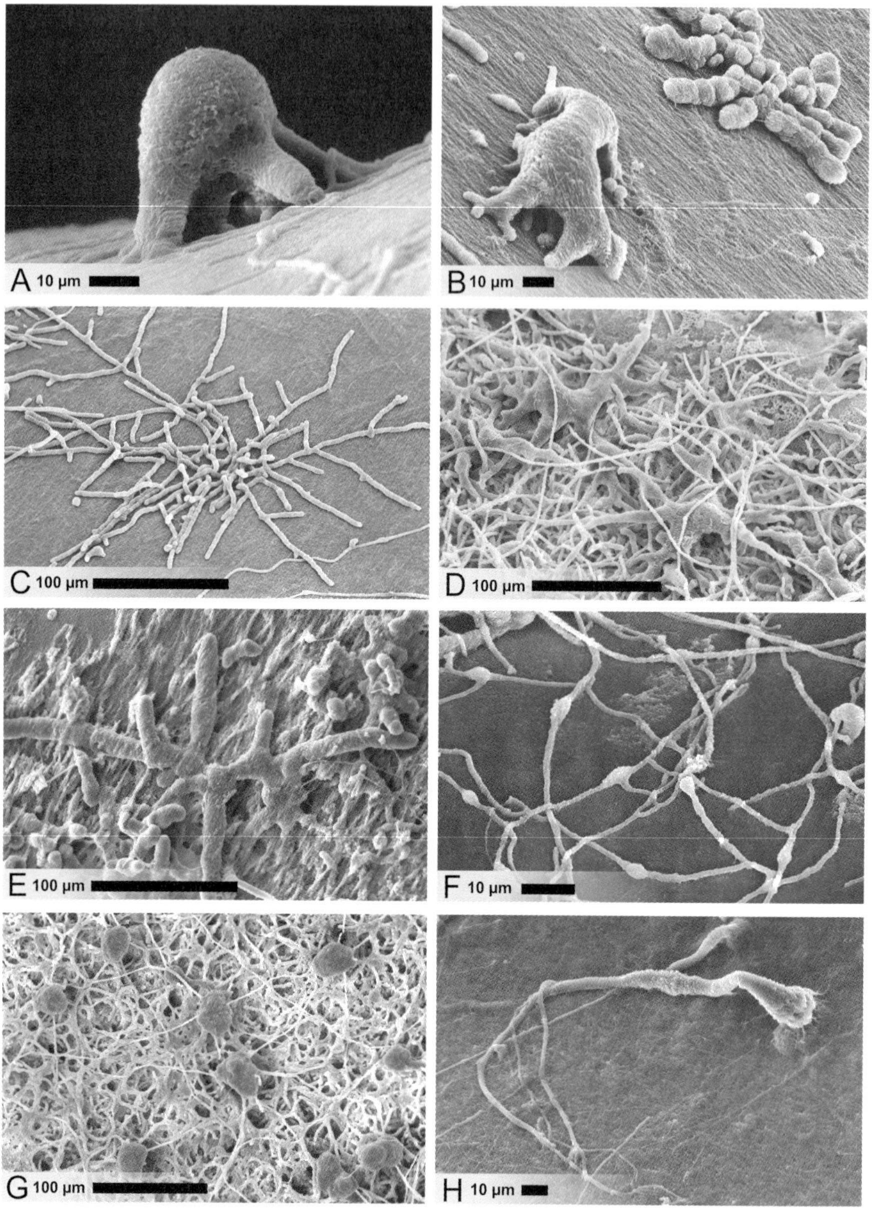

Fig. 32 Traces produced by chlorophytes (**A**-**D**) and fungi (**E**-**H**) recorded in *Littorina* shells collected at the shoreline (SEM): **A** Lateral view of *Cavernula pediculata*. **B** Large, wide variation of *C. pediculata* together with *Fascichnus*. **C** Feather-like branching pattern of *Eurygonum pennaforme*. **D** Dense *Ichnoreticulina elegans* carpet with large dendritic cavities. **E** Rectangular branching pattern of *Orthogonum lineare*. **F** Thin *Orthogonum fusiferum* with spindle-shaped swellings. **G** *Saccomorpha clava* towering a dense microboring layer. **H** Collapsed deeply boring '*Flagrichnus*-form 1'

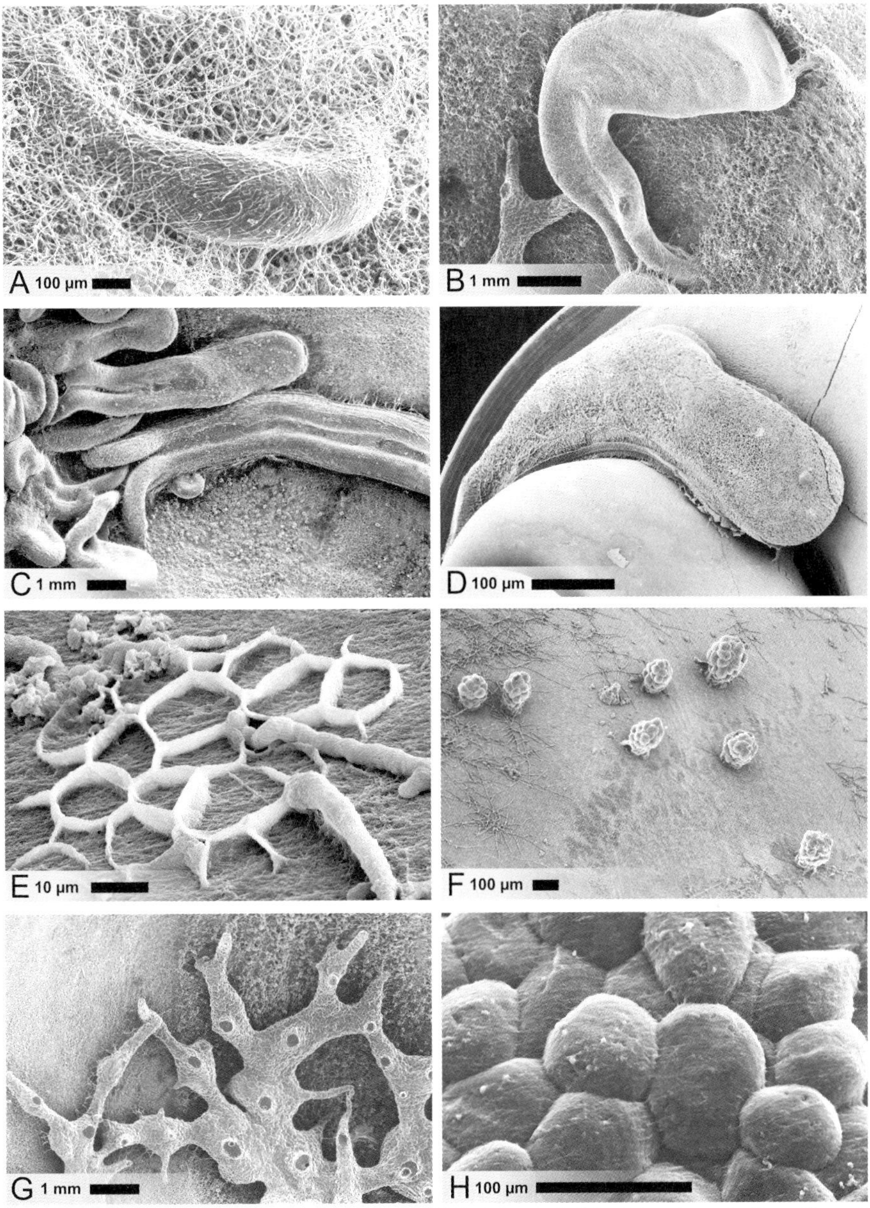

Fig. 33 Traces produced by macroborers (SEM) recorded in *Littorina* shells collected at the shoreline: **A** Initial *Caulostrepsis*. **B** Mature *Caulostrepsis taeniola*. **C** Densely-spaced *C. taeniola* showing affinity to *Maeandropolydora* where proximal galleries diverge and to *C. contorta* where borings merge. **D** Proximal part of *C. cretacea* without distinct vane. **E** Very initial boring of a clionaid sponge with first boring cells penetrating into the substrate. **F** Initial raspberry-shaped cavities of *Entobia*. **G** Mature multicamerate network of *Entobia*. **H** Close-up showing the characteristic verrucose micro-sculpture

4.5 Spatial and temporal patterns of bioerosion

The experimental approach of this study provided the opportunity to observe the temporal evolution of colonisation by microborers in relation to light availability and bathymetry (Figs. 34-35).

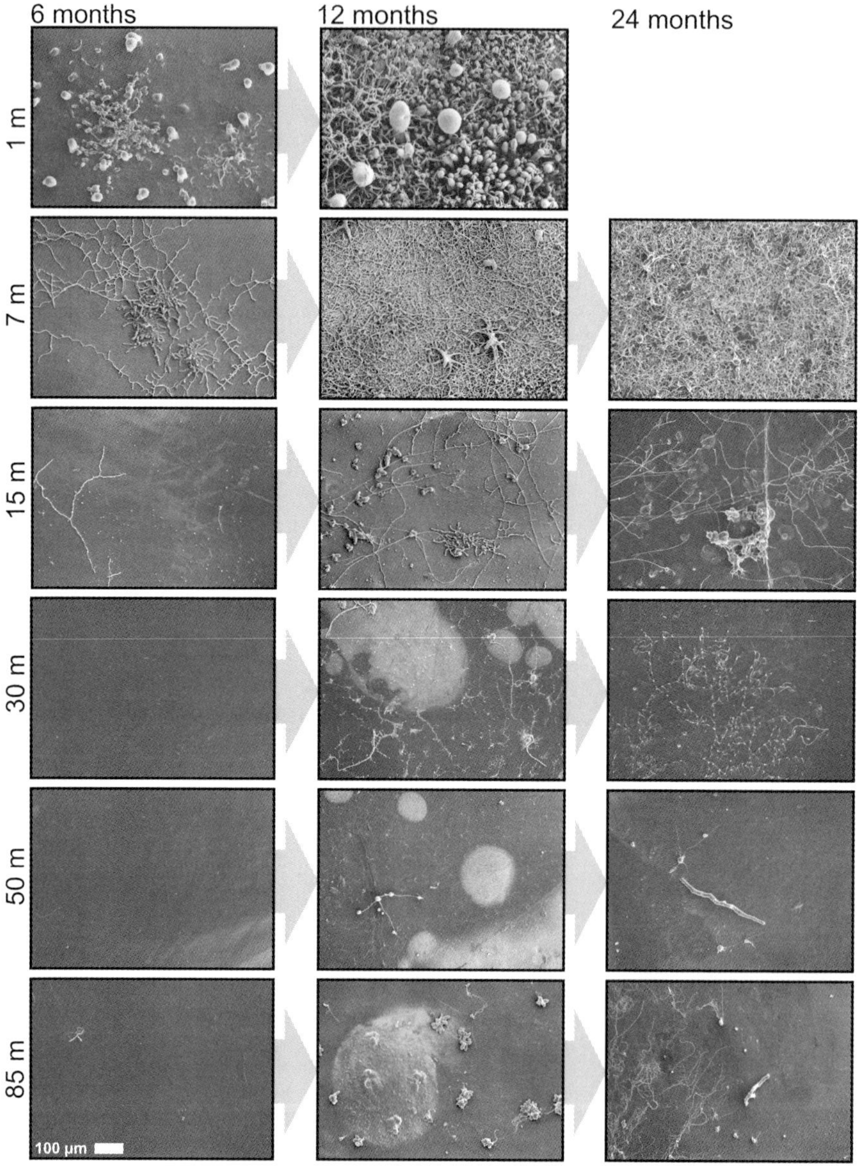

Fig. 34 The development of the ichnocoenoses in shells of the bivalve *Callista chione* at the various depth stations after 6, 12 and 24 months exposure

In the following account, the development of the ichnocoenoses at each depth as reflected in the SEM casts is outlined. Subsequently, the development of the corresponding biocoenoses studied via light-microscopic analyses of the Iceland spars and bivalve shells are characterised. Finally, these results are compared to the ichnocoenoses encountered in the taphonomically advanced background samples.

Ichnocoenoses development

After being exposed for just 6 months, a considerable number of traces were encountered at the 1 m station, dominated by large quantities of *Cavernula pediculata* and various taxa of *Fascichnus*. The (juvenile) traces of *Cavernula pediculata* show penetration depths of up to 50 µm. Individual *Fascichnus dactylus* and *Fascichnus frutex* colonies reach diameters of up to 1.5 mm at a penetration depth approaching 100 µm. After 12 months exposure, the substrate was densely colonised by microborers, still dominated by *Cavernula pediculata* and various taxa of *Fascichnus*. In addition, a number of further ichnotaxa were scarcely encountered: *Planobola*, *Ichnoreticulina elegans*, *Rhopalia catenata*, *Scolecia filosa*, *Eurygonum nodosum* and *Eurygonum pennaforme*. The chlorophyte/cyanobacteria-dominated ichnocoenosis encountered after 1 year exposure comprises 10 ichnotaxa and can already be considered as mature, in that no significant changes in the established microboring community are expected to take place with progressing exposure time. At the same depth (1 m) but in the sheltered lagoonal setting, the rate of colonisation was even higher. After 6 months exposure, the substrates were densely colonised by microborers. The ichnocoenosis is dominated by *Cavernula pediculata*, *Eurygonum nodosum*, *Fascichnus dactylus* and *Orthogonum fusiferum*. After 18 months, the same composition, but with a stronger dominance of *Eurygonum nodosum* and *Orthogonum fusiferum* was found. Additionally, *Planobola* and *Fascichnus* cf. *acinosus* were encountered. By this time, the substrate was bored in several tiers to a depth of several 100 µm (deepest penetration by *Orthogonum fusiferum*). However, with only 5 ichnotaxa, the diversity is low when compared to the corresponding depth at the transect.

Only few metres deeper, at the 7 m station, a very different, although comparably diverse ichnocoenosis was encountered. After 6 months exposure, the substrate was colonised by numerous colonies of *Ichnoreticulina elegans*, '*Fascichnus*-form 1', *Eurygonum pennaforme* and *Eurygonum nodosum*, with individual colonies occasionally exceeding 1 mm in diameter. Additionally, *Fascichnus dactylus*, *Fascichnus frutex* and the shallowest records of '*Flagrichnus*-form 1' and '*Flagrichnus*-form 2' were observed.

After 12 months exposure, *Ichnoreticulina elegans* densely covered the casts in multiple tiers partly masking shallower layers of the ichnocoenosis. *Scolecia filosa* and *Cavernula pediculata* were rare further constituents of this ichnocoenosis and after 2 years exposure, the ichnocoenosis was complemented by *Rhopalia catenata*. The dominant trace makers (like for the 1 m station) were the cyanobacteria and chlorophytes, representing the majority of the 12 ichnotaxa identified.

With the vanishing light at the 15 m station, the pace of ichnocoenosis development is greatly reduced. After 6 months exposure, only few small colonies of *Ichnoreticulina elegans* were encountered, besides rare specimens of *Rhopalia catenata* and '*Fascichnus*-form 1'. At this depth and downwards, '*Flagrichnus*-form 1' and '*Flagrichnus*-form 2' were the dominant traces, clustered in patches of hundreds of individuals. This dominance became even more pronounced after 12 and 24 months exposure. *Scolecia filosa, Orthogonum lineare* and '*Orthogonum*-form 2' were further rare constituents of this ichnocoenosis of medium to high diversity (9 ichnotaxa).

At 30 m and downwards, only traces of heterotrophic organisms were encountered, except for very few initial colonies of *Ichnoreticulina elegans* in the 2 year samples. After 6 months exposure the substrates were still almost pristine and only '*Flagrichnus*-form 1' and '*Flagrichnus*-form 2' were present in small numbers. All other traces made their appearance not before one year of exposure, such as few small and only rarely linked sporangia of *Saccomorpha clava* and short stretches of *Orthogonum lineare*. Both became more abundant after two years, but with *Orthogonum lineare* galleries seldomly exceeding 500 μm in length and only few branching points. Rare traces were *Saccomorpha terminalis*, *Planobola radicatus*, '*Orthogonum*-form 1', and *Scolecia serrata*. At 85 m, the highest diversity (7 ichnotaxa) of the aphotic stations (50 m: 5 ichnotaxa) was found.

In summary, the highest diversity is observed at 7 m water depth with 12 ichnotaxa produced by microendoliths, followed by the 1 m station (10 ichnotaxa; Fig. 35). The lagoonal setting is considerably depleted in ichnotaxa richness (only 5 ichnotaxa). With the vanishing light availability towards greater depths, the diversity decreases to only 5 ichnotaxa at the 50 m with a slight increase at the vicinity of the cold-water coral reef (7 ichnotaxa). A corresponding pattern is developed in the case of macroborings. However, only at the shallow stations the ichnocoenoses can be considered mature after just two years exposure. This is especially true with respect to the contribution by macroborers, which can be expected to gain importance with progressing substrate degradation as indicated by the taphonomically advanced background samples .

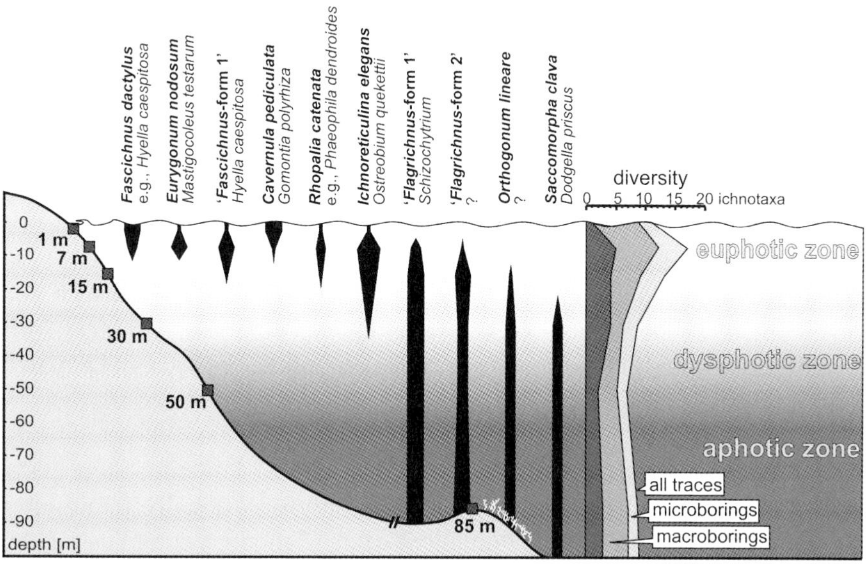

Fig. 35 The bathymetric transect in the northern Kosterfjord area in relation to the photic zonation with the position of the different depth stations (left), the bathymetric distribution of the most important ichnotaxa, and the diversity trend of microborers and macroborers (right)

The degree of bioerosion is highest at the shallowest stations, especially in the lagoonal setting, and is continuously decreasing with the availability of light towards aphotic depths (Fig. 34). There, the diversity and progress of bioerosion is determined either by hydrographic parameters such as current velocities and/or by the proximity to the cold-water coral reef with its generally enhanced biodiversity (Jonsson et al. 2004).

Microendoliths in Iceland spars and mollusc shells

The composition of the ichnocoenoses is generally confirmed by the distribution pattern of the living endoliths. The diversity is accordingly highest at 7 m (11 species after 12 months exposure) closely followed by the two 1 m settings (10 species after 6 months exposure). At both depths the highest diversity is shown by the cyanobacteria (7 and 6 species, respectively). The diversity drops significantly at 15 m (6 species after 12 months) and downwards. Below 30 m, only few fungi were encountered (the organic bearing shell material from 24 months exposure was not analysed, but would have probably increased the record of fungi analogue to the corresponding epoxy resin casts).

The two shallow-water stations (1 m) allow a comparison of a more open marine habitat against a lagoonal setting. The endolithic species in the lagoonal station were dominated by cyanobacteria. After 6 months exposure, 6 species occurred in Iceland spars and shells, while only 4 chlorophytes colonised the substrates. At the same time, very few endoliths were observed in the open-water habitat where only a single chlorophyte and one fungal species were recorded. *Gomontia polyrhiza* occurred in very high abundance in both settings. This picture reversed after one year exposure when a total of 7 cyanobacteria and 3 chlorophytes were encountered at the transect. The development in the lagoon after 18 months exposure shows a clear shift towards chlorophytes (4 species) and fungal endoliths (2 species). Only two cyanobacteria occurred, *Hyella caespitosa* and *Mastigocoleus testarum*.

At the 7 m station, chlorophytes dominated the biocoenosis already after 6 months exposure with *Gomontia polyrhiza* covering large parts of the substrate. *Ostreobium quekettii* was only occasionally found after 6 months exposure but clearly dominated after 12 months, complicating the identification of other taxa within the dense carpets of this siphonal alga. The cyanobacteria *Hyella balani* and *Hyella gigas* were found frequently besides small numbers of *Mastigocoleus testarum*, *Plectonema terebrans*, *Hyella caespitosa*, *Solentia achromatica* and the chlorophytes *Phaeophila dendroides* and *Eugomontia sacculata*.

At 15 m only a few cyanobacteria remained active. *Plectonema terebrans*, *Hyella caespitosa, Hyella gigas, Solentia achromatica* and *Mastigocoleus testarum* were found together with the dominating *Ostreobium quekettii*.

Below 30 m, exclusively scarcely distributed heterotrophic endoliths were encountered such as the fungi *Phytophthora* and *Dodgella priscus*.

Comparing experimental substrates and background samples

When comparing the very initial ichnocoenoses encountered at the 85 m Säcken Reef station and the mature ichnocoenoses recorded in the background material from the same site (Sect. 4.3), the comparatively large number of macroboring traces in the background samples is evident. While sponge borings were limited to few initial *Entobia* cavities in the experimental substrates, this ichnotaxon represents the dominant trace in the *Lophelia* skeletons, complemented by half a dozen rare potential micro-sponge traces ('Microsponge forms 1-6'). The brachiopod attachment scar *Podichnus centrifugalis*, the foraminiferal trace 'Foraminiferan-form 3', the two problematic traces and '*Orthogonum*-form 3' were exclusively encountered in the background sample material but were of no quantitative relevance. Traces produced by boring bryozoans, especially the ichnogenus

Pennatichnus, which is an important compound of the ichnocoenosis in the *Lophelia* skeletons, was found to be missing in the experimental substrates altogether. In terms of fungi, *Saccomorpha clava* and *Orthogonum lineare* – the key ichnospecies for the aphotic index ichnocoenosis – were present in considerable numbers in both substrates, while the two '*Flagrichnus*-forms' and *Saccomorpha terminalis* were in turn only encountered in the experimental substrates. This fact may be explained by the latter traces possibly being produced by opportunistic borers while infestation by *Dodgella priscus* and the unknown producer of *Orthogonum lineare* take over during advanced taphonomic stages. The delayed appearance of macroborers is a typical pattern and confirms the observations of previous bioerosion experiments (e.g., Kiene & Hutchings 1994b).

In contrast to the aphotic end member, the microborer ichnocoenosis established in the shallow-water experimental substrates can be regarded as mature. Except for *Fascichnus rogus*, all ichnotaxa recorded in the *Littorina* shells were also found in the experimental substrates as an intersection of the 1 m and 7 m stations. The dominant ichnotaxa are corresponding at both sample sets. Differences are expressed by the relative scarcity of *Eurygonum nodosum*, '*Fascichnus*-form 1' and *Ichnoreticulina elegans* as well as a superior abundance of *Scolecia filosa* and *Eurygonum pennaforme* in the *Littorina* shells. A number of traces were in turn exclusively encountered in the experimental substrates but only in minute quantities, as there are *Fascichnus* cf. *acinosus*, *Rhopalia catenata* and '*Flagrichnus*-form 2'. Such as for the *Lophelia* skeletons, the macroborers (boring sponges and polychaetes) are more prominent in the advanced stages of shell degradation. In conclusion, the results obtained from the experimental substrates mirror the ones recorded from the background *Littorina* samples quite well even though the latter being a mobile substrate of different shape and ultrastructure. The composition of the ichnocoenoses with components from both, the 1 m and 7 m experimental substrates reflects the bathymetrically restricted habitat of *Littorina* in the euphotic uppermost metres of the water column.

4.6 The ichnocoenoses in relation to bathymetry

When applying the established scheme (see Sect. 1.7) to the present material, the general trends in bathymetric distribution of the various trace maker groups and their corresponding traces can be clearly confirmed (Fig. 35). The individual index ichnocoenoses, however, were of limited applicability, except for the aphotic *Saccomorpha clava* / *Orthogonum lineare* ichnocoenosis. This may partly be due to the considerable fluctuations of temperature and salinity in the shallow waters hindering the development of the defined shallow-water index ichnocoenoses. The key-ichnospecies *Fascichnus acinosus* (shallow euphotic zone II), is very (too) rare in the present material and the rhodophyte trace '*Palaeoconchocelis starmachii*' (shallow euphotic zone III and deep euphotic zone) is missing altogether. The latter is not a clearly defined ichnotaxon (and thus written here in quotation marks), questioning its suitability as a key-ichnospecies. *Cavernula pediculata* has a cosmopolitan biogeographic distribution and is frequently found in most shallow-marine settings and is thus suggested as a substitute for the even in the tropics rare ichnospecies *Fascichnus acinosus*. For '*Palaeoconchocelis starmachii*', a suitable substitute is given by '*Fascichnus*-form 1' (in open nomenclature as yet), respectively. The dysphotic ichnocoenosis is clearly indicated in the present setting by *Ichnoreticulina elegans* being the only trace produced by a phototrophic organism.

Due to the high northern latitude (59°N) and a considerable eutrophication, the photic zonation in the study area is highly condensed when compared to tropical and subtropical environments, where the boundaries of the various photic zones are located much deeper (deepest occurrence of algae: 268 m; Littler et al. 1985). Only the 1 m stations can be attributed to the shallow euphotic zone (II-III), whereas the 7 m must be considered as deep euphotic. The 15 m station marks the transition to the dysphotic zone and from 50 m downwards only heterotrophic organisms are encountered. The ichnospecies *Ichnoreticulina elegans* and/or its trace maker – the green alga *Ostreobium quekettii* – has been shown to be a well-suited indicator for the photic limit due to its high abundance in most marine settings and its ability to cope with extremely low light availabilities (Akpan & Farrow 1984a and b; Akpan 1986). In the present setting, the deepest occurrence of *Ichnoreticulina elegans* was encountered at the 30 m station (limit of all other obligate phototrophs: 15 m), where it appeared only sporadically after 2 years exposure, and the shallowest non-occurrence at 50 m. These results are well confirmed by the direct determinations of the photic boundaries based on the measurements of the PAR (Figs. 9, 35; see above).

The results agree with the determination of the photic limit (determined by the deepest occurrence of *Ostreobium quekettii* and rhodophyte *Conchocelis*-stages) in northern Scottish coastal waters, located at a comparable northern latitude (56°-59°N) by Akpan & Farrow (1984a), with a range from 40 m for the Orkney shelf to about 22 m for the partly enclosed firths. Glaub et al. (2002), applying ichnocoenoses analysis on the same samples, judged the base of the euphotic zone to be at about 20 m water depth and additionally report 16 m for the Tromsø area (70°N, Norway).

5 Carbonate accretion patterns

The study of marine hardground communities in Archaean to Recent settings support a vast literature (see Taylor & Wilson 2003 for a recent review). From the actuo-palaeontologic perspective, sessile calcareous organisms are of special interest among the various benthic biota settling on hard substrates since they yield superior potential of becoming part of the fossil record in alliance with the traces left by boring organisms.

5.1 The carbonate accretion inventory

Calcareous epizoans found on the experimental substrates comprise serpulimorphs, bryozoans, balanids, crinoids and epibenthic foraminiferans. While the former 4 groups were only briefly investigated and point-counted on family level, the foraminiferans as the most abundant and diverse group (which also contributes to bioerosion) were quantified on species level and subjected to an in-depth study. A complete data matrix of the point-counting quantification is found in the Appendix 1 (serpulimorphs, bryozoans, balanids, crinoids) and Appendix 2 (foraminiferans).

Serpulimorphs

Serpulimorphs were among the first to settle on the experimental plates in large numbers. The two families present are the Serpulidae such as *Filograna* and *Hydroides* (Fig. 36A), and the Spirorbidae such as *Spirorbis* (Fig. 36B). The by far dominating species are *Hydroides norvegica* and *Spirorbis spirorbis*.

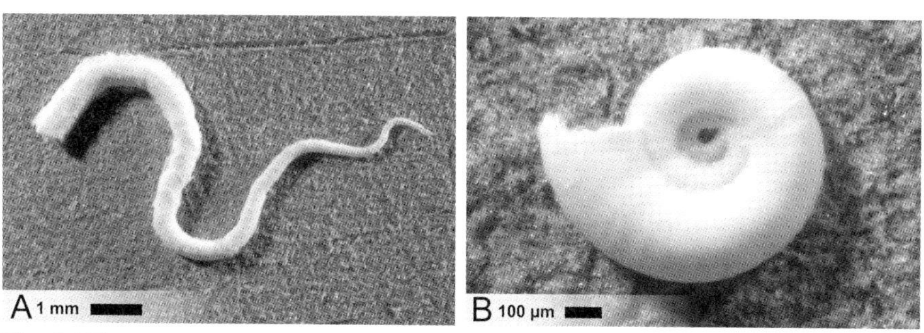

Fig. 36 The most common representatives of the two polychaete families Serpulidae – *Hydroides norvegica* (**A**) and Spirorbidae – *Spirorbis spirorbis* (**B**)

Bryozoans

Among the most prominent calcareous epizoans encountered on the experimental substrates are the cyclostome and cheilostome bryozoans and specifically representatives of the families Tubuliporidae and Escharellidae, respectively. The two most abundant species *Tubulipora liliacea* (Fig. 37A-B) and *Escharella immersa* (Fig. 37C-D) were found in various stages of colony development ranging from an ancestrula with very few initial zooids to large colonies several centimetres in diameter. All bryozoans were distinctly more abundant on the bottom side of the experimental frames than on the top side.

Fig. 37 The two most abundant bryozoans encountered on the experimental substrates: **A** Juvenile colony of *Tubulipora liliacea*. **B** Delicate adult colony of *Tubulipora liliacea*. **C** Ancestrula with one further zooid of *Escharella immersa*. **D** Adult colony of *Escharella immersa*

Balanids

Especially in the shallowest waters, balanids are ubiquitous hardground dwellers and were also encountered in large numbers on the experimental substrates. By far the most abundant species recorded was *Balanus improvisus* (Fig. 38A), which formed a dense carpet on the substrates at 1 m water depth already after 12 months exposure. Besides these Balanidae, only one specimen of a representative of the familiy Verrucidae was encountered, namely *Verruca stroemia* (Fig. 38B).

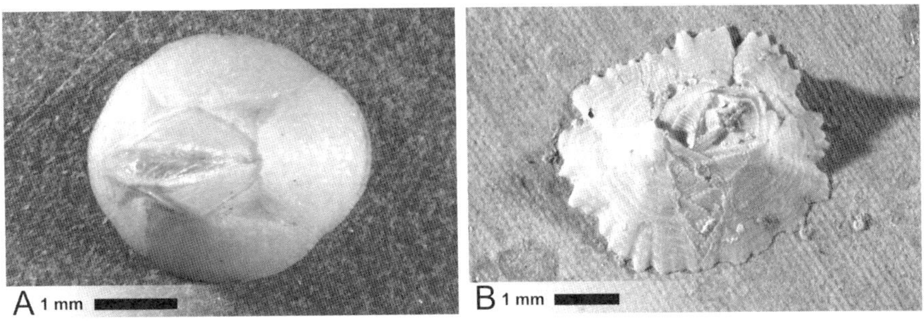

Fig. 38 The by far dominating balanid representative *Balanus improvisus* (**A**) and the only record of the verrucid *Verruca stroemia* (**B**)

Crinoids

A rare finding were star-shaped crinoid basal plates (Fig. 39A), less than half a millimetre in diameter, belonging to the family Antedonidae. SEM imagery of the skeletal ultrastructure reveal the three-dimensional calcite architecture typical for echinoderms (Fig. 39B). They were encountered already after 6 months exposure and exclusively at the 85 m station at the Säcken Reef site.

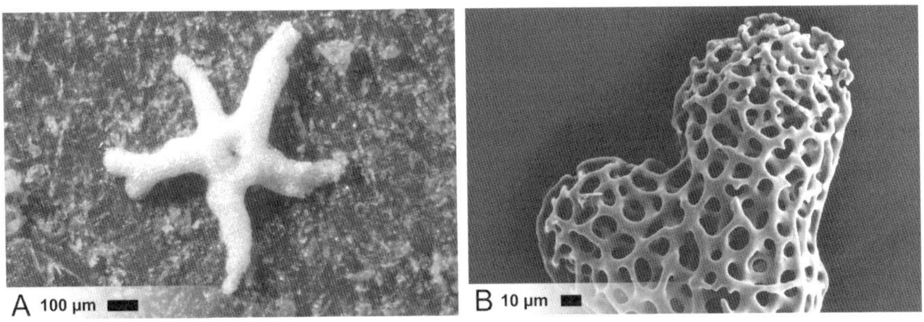

Fig. 39 Star-shaped rhizoidal parts of a crinoid (**A**) and a close-up (SEM image), revealing the echinoderm skeletal ultrastructure (**B**)

Foraminiferans

The benthic foraminiferal assemblage settled on the investigated artificial substrates comprises 12 different species, 8 of which belonging to the Rotaliina and 4 to the Textulariina (Table 9; Fig. 40).

The most abundant species belonging to the Rotaliina are *Cibicides lobatulus* (51.2% of the total foraminiferal assemblage), *Nubecularia lucifuga* (6.5%) and *Planorbulina mediterranensis* (3.2%). All other rotaliids (*Rosalina anomala*, *Epistominella vitrea*, *Gypsina vesicularis*, *Elphidium incertum*, *Cassidulina obtusa*) show an accessory contribution to the assemblage with <0.12% in relative abundance. The dominating calcareous species *Cibicides lobatulus* represents mean 83.5% of all rotaliniid species. Species belonging to Textulariina are *Lituotuba lituiformis*, *Tholosina vesicularis*, *Textularia truncata*, and *Tritaxis fusca*. *Lituotuba lituiformis* is the dominant agglutinating form with a relative abundance of 37.5% and describes 96.9% of all agglutinating species. *Tholosina vesicularis* is the second dominant species with 2.9%, while *Textularia truncata* and *Tritaxis fusca* only contribute 0.2% to the relative abundance of agglutinating species.

Table 9 Benthic foraminiferans settling on investigated artificial substrates (PVC + carbonate plates) with their relative abundance and bathymetric distribution (for a faunal reference list, please see Wisshak & Rüggeberg 2006)

Species	Relative abundance [%]	Bathymetric distribution [m]
Cibicides lobatulus (Walker & Jacob, 1798)	51.24	7-85
Lituotuba lituiformis (Brady, 1879)	37.51	7-85
Nubecularia lucifuga Defrance, 1825	6.52	7-15
Planorbulina mediterranensis d'Orbigny, 1826	3.24	7-85
Tholosina vesicularis (Brady, 1879)	1.11	7-50
Rosalina anomala Terquem, 1875	0.12	15-85
Epistominella vitrea (Parker, 1953)	0.11	7-30
Gypsina vesicularis (Parker & Jones, 1860)	0.06	15
Textularia truncata Höglund, 1947	0.06	30-85
Elphidium incertum (Williamson, 1858)	0.05	15-30
Tritaxis fusca (Williamson, 1858)	0.02	7-15
Cassidulina obtusa (Williamson, 1858)	0.01	50

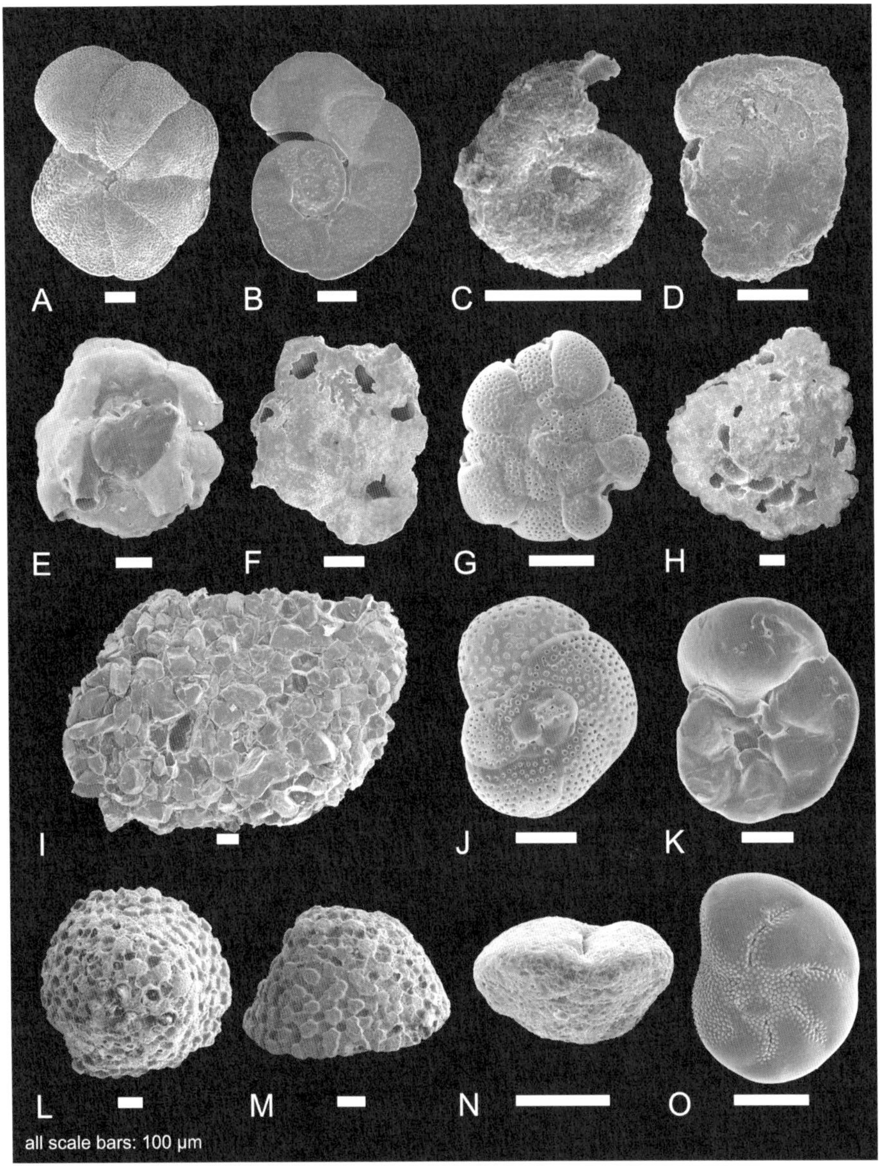

Fig. 40 Selected benthic foraminiferans (in order of decreasing abundance) found attached on the artificial substrates (SEM images): **A-B** Dorsal and ventral view of *Cibicides lobatulus*. **C-D** Dorsal and ventral view of *Lituotuba lituiformis*. **E-F** Dorsal and ventral view of *Nubecularia lucifuga*. **G-H** Dorsal and ventral view of *Planorbulina mediterranensis*. **I** Dorsal view of *Tholosina vesicularis*. **J-K** Dorsal and ventral view of *Rosalina anomala*. **L-M** Dorsal and lateral view of *Gypsina vesicularis*. **N** Dorsal view of *Textularia truncata*. **O** Dorsal view of *Elphidium incertum*

5.2 Bathymetric distribution and diversity

Analogue to the bioerosion agents, the calcareous epibenthos is not evenly distributed with depth and time but shows characteristic distribution patterns, controlled by environmental factors such as watermass properties, hydrodynamic force or competition for space and/or food among different biota.

The overall means of the bathymetric distribution of the various epizoans for all exposure times and substrate types (Fig. 41) clearly express the dominance of the small epibenthic foraminiferans in terms of population densities. Specifically *Cibicides lobatulus* and *Lituotuba lituiformis* by far outnumber any other epizoans by reaching population densities of many thousands per square metre. Among the larger epizoans, solely the balanids reach comparable numbers, but only at the shallowest station where they outcompete all other epizoans. In addition, the serpulimorphs appear in considerable numbers, while the bryozoans are comparatively rare and crinoid rhizoidal parts were only found at the deepest station. While most groups and especially the foraminiferans show their distribution maximum in the shallower waters in 7 m to 15 m water depth, others increase in abundance towards deeper waters, such as the Spirorbidae, the Tubuliporidae or the Antedonidae. Concerning the overall diversity, the number of non-foraminiferal epizoan families slightly increases with depth while the number of foraminiferal species decreases (Table 9).

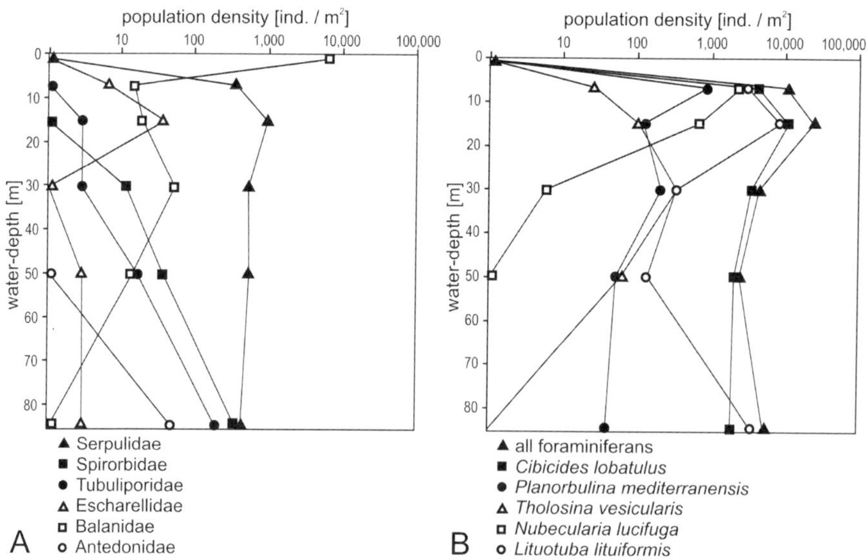

Fig. 41 A logarithmic plot of the mean abundance of the various epizoans with depth on all substrate types and exposure times for the serpulimorphs, bryozoans, balanids and crinoids (**A**) and for the foraminiferans (**B**)

When considering the different exposure times in the bathymetric distribution and when focusing solely on the PVC settlement plates, the patterns become more heterogeneous (Fig. 42). Only few groups such as the two bryozoan families Tubuliporidae and Escharellidae draw a steady increase in specimen numbers with time and with distinct bathymetric maxima at 85 m and 15 m, respectively. Others decrease with time (Antedonidae and Spirorbidae) or show a completely heterogeneous pattern as for instance the Serpulidae. Among all recorded families, only the Balanidae are present at the shallowest station (1 m) with high numbers exceeding 10,000 ind./m^2 after 12 months exposure (24 months panel lost).

A more consistent picture is drawn by the foraminiferans. The combined overall bathymetric distribution of the five main species (*Cibicides lobatulus*, *Planorbulina mediterranensis*, *Nubecularia lucifuga*, *Lituotuba lituiformis* and *Tholosina vesicularis*) shows a pronounced maximum at the 15 m station with a specimen density of close to 80,000 ind./m^2 after two years exposure (Fig. 43A). This abundance pattern applies for all three exposure periods (6, 12 and 24 months) with a strong increase in specimen numbers with time. The relative abundance after 6 months exposure is minute for all five species. The specimen densities significantly drop below 30,000 ind./m^2 towards shallower as well as deeper waters but slightly increase at the deepest station (85 m). At the shallowest station in 1 m water depth, no foraminiferans were recorded after 6 and 12 months exposure (24 months panel lost).

Even though this overall bathymetric distribution pattern is generally reflected also by the individual species (Fig. 43B-F) and especially by *Cibicides lobatulus* (Fig. 43B), which accounts for more than 50% of the total number of specimen (Table 9), certain individual patterns emerge. *Planorbulina mediterranensis* shows a clear distribution maximum at only 7 m water depth and an overall decrease with depth (Fig. 43C). *Nubecularia lucifuga* is approximately as abundant at 7 m as 15 m, but is very rare at 30 m and absent below that depth (Fig. 43D). While all three species show a strong increase in specimen numbers with exposure time, *Lituotuba lituiformis*, the second most abundant species, shows near-constant densities after 12 and 24 months exposure. Besides a maximum at 15 m water depth, this species has a second peak at 85 m (Fig. 43E). *Tholosina vesicularis* was found with a more irregular distribution pattern, exhibiting a maximum at 30 m after 12 months exposure and a considerably lower peak at 15 m after 24 months exposure (Fig. 43F). *Tholosina vesicularis* was absent at 85 m.

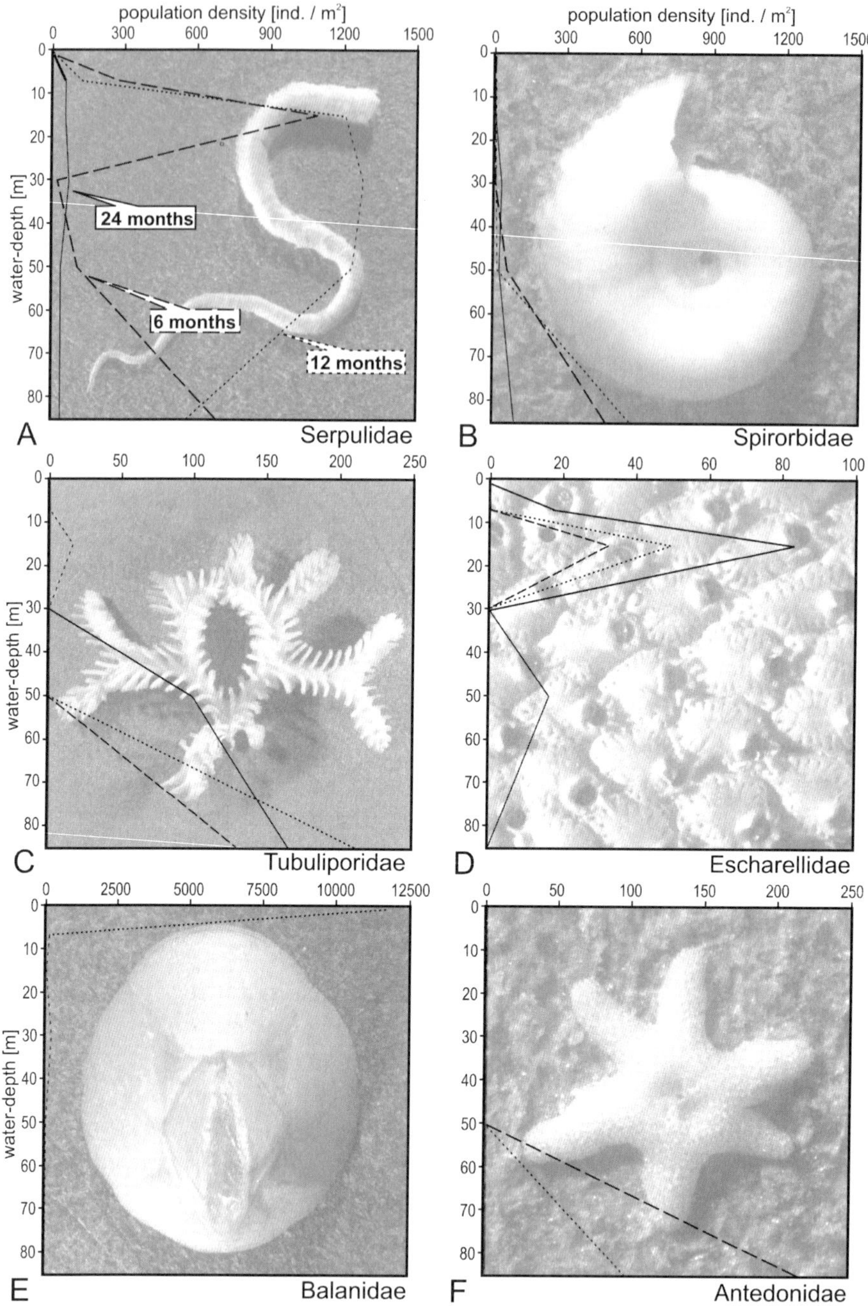

Fig. 42 Bathymetric distribution of the Serpulidae (**A**), Spirorbidae (**B**), Tubuliporidae (**C**), Escharellidae (**D**), Balanidae (**E**) and Antedonidae (**F**) after 6, 12 and 24 months of exposure on the PVC plates (note the different scaling of the x-axes)

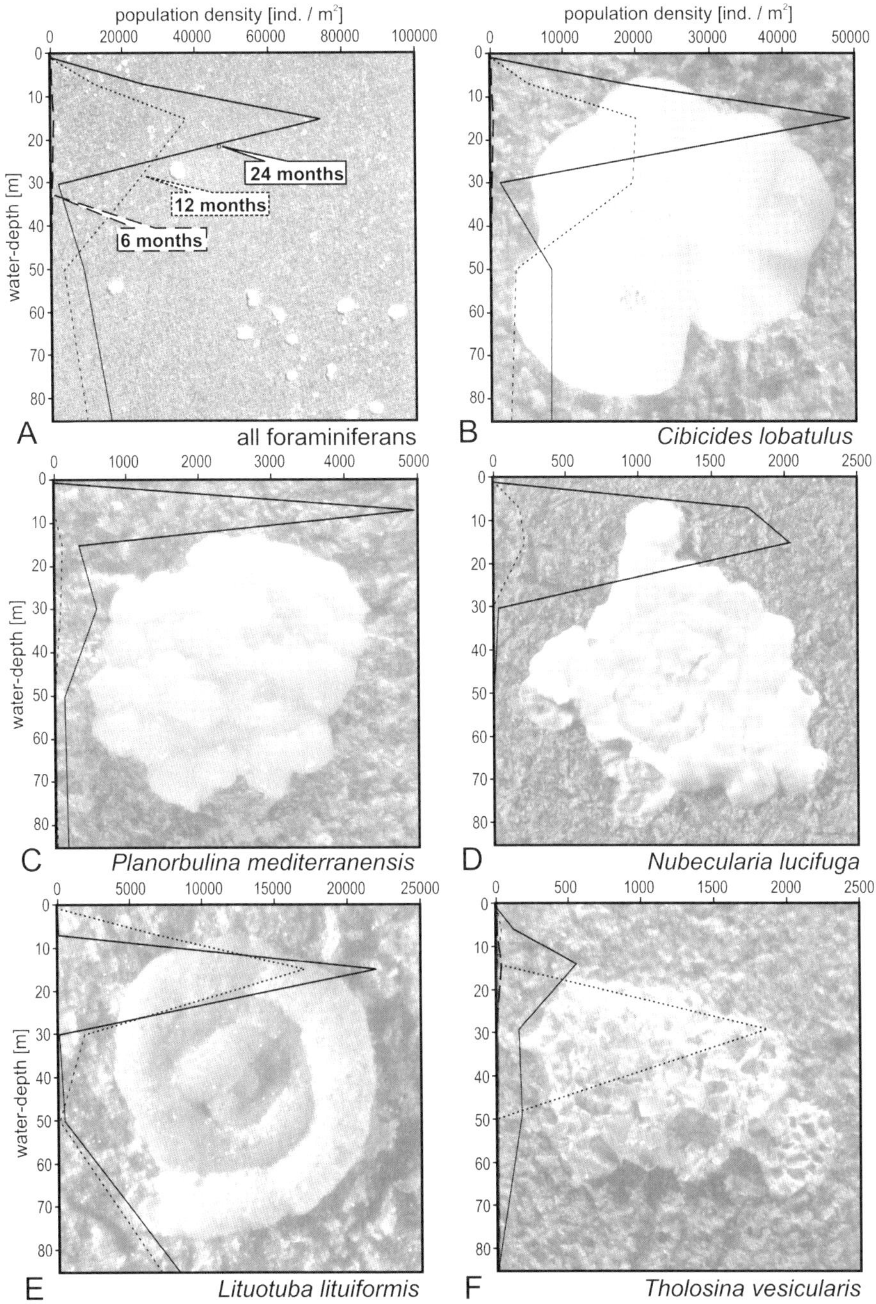

Fig. 43 Bathymetric distribution of all foraminiferans (**A**), *Cibicides lobatulus* (**B**), *Planorbulina mediterranensis* (**C**), *Nubecularia lucifuga* (**D**), *Lituotuba lituiformis* (**E**) and *Tholosina vesicularis* (**F**) after 6, 12 and 24 months exposure on the PVC plates, yielding abundance maxima at 7 m and 15 m water depth (note the different scalings of the x-axes)

The species composition at the various depth stations and exposure times as recorded on the PVC settlement plates (Fig. 44) is largely dominated by *Cibicides lobatulus* and *Lituotuba lituiformis*. *Cibicides lobatulus* is the most abundant species and is only outnumbered by *Lituotuba lituiformis* at 7 m after 12 months and at 85 m after 12 and 24 months exposure (Fig. 44B-C). With few exceptions (*Nubecularia lucifuga* at 30 m after 12 months; *Nubecularia lucifuga* and *Planorbulina mediterranensis* at 30 m after 24 months exposure), all other species are subordinate members of the foraminiferal assemblage. The absolute dominance of *Cibicides lobatulus* is most clearly expressed at the 30 m and 50 m stations (Fig. 44D) and at all depths after 6 months exposure, where, however, only a limited total number of individuals was recorded (Fig. 44A).

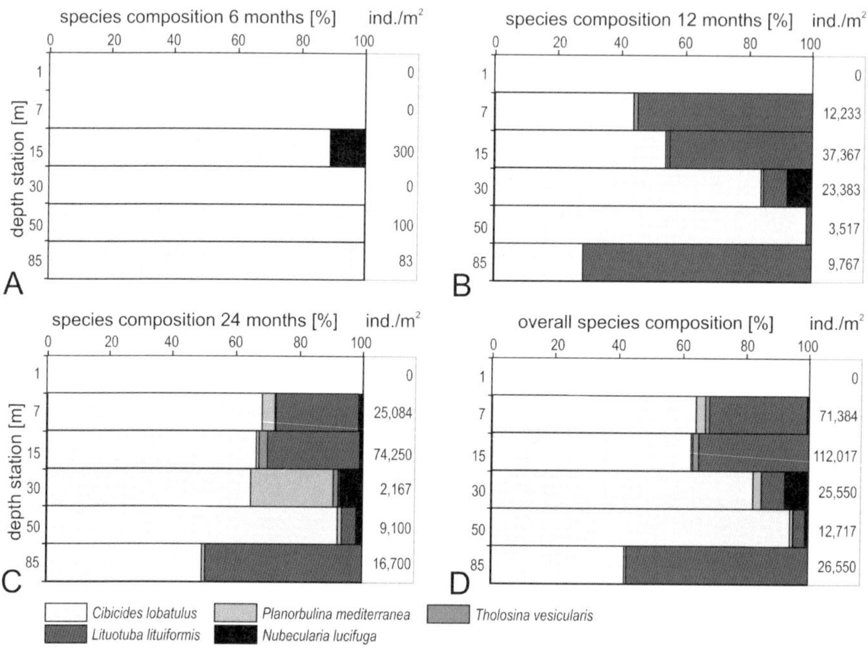

Fig. 44 The cumulative relative abundance of the dominant taxa on the PVC plates and the extrapolated number of individuals per m² at the various depth stations after 6 (**A**), 12 (**B**) and 24 months (**C**) and arithmetic means of all exposure times (**D**)

The maximum species diversity is encountered at the same depth than the maximum specimen density. The highest diversity is found at 15 m and 30 m water depth with 10 and 9 species, respectively (see Table 9). At 7 m and 50 m moderate 7 species and at 85 m only 5 species are present.

5.3 Substrate preference

When comparing the population densities of the various epizoans on the carbonate versus the PVC plates, specific substrate preferences become apparent (Fig. 45). The Serpulidae are moderately more abundant on the carbonate substrates at all water depths while the Spirorbidae show a comparable distribution on both substrate types. For the bryozoans, the Escharellidae are more abundant on the PVC plates at most depths while the Tubuliporidae yield an inconsistent pattern. The balanids show an only moderate preference for the carbonate plates and the crinoids are more abundant on the PVC substrates.

The comparison of the population densities on the carbonate versus the PVC plates yields an almost threefold abundance of foraminiferal tests on the latter (Fig. 46). The difference becomes even more prominent when considering individual species. *Cibicides lobatulus* is 25 times more abundant on the PVC plates. For *Planorbulina mediterranensis*, *Tholosina vesicularis* and *Lituotuba lituiformis*, the difference is less pronounced. *Nubecularia lucifuga* on the other hand is about 4 times more abundant on the carbonate plates. The distribution pattern within one substrate type and even within individual settlement plates is heterogeneous. Some species and especially *Lituotuba lituiformis* often appear in distinct clusters, while others, such as *Cibicides lobatulus* and *Planorbulina mediterranensis*, are more evenly dispersed.

5.4 Discussion of the foraminiferal assemblage

Foraminiferal assemblage

Most of the benthic foraminiferal species encountered in the Kosterfjord experiment have also been reported from other Scandinavian fjord settings (for example Alve & Nagy 1986; Austin & Sejrup 1994; Gustafsson & Nordberg 1999; Klitgaard-Kristensen & Buhl-Mortensen 1999). Only the second-most abundant calcareous species *Nubecularia lucifuga* and *Gypsina vesicularis* lack subrecent to fossil occurrence in surface sediment samples, which may be due to breakage and dissolution of the tests. *Nubecularia lucifuga* thrives in shallow waters (<15 m water depth) in the Spencer Gulf, Australia (Cann et al. 2000) and seems to be correlated with shallow subtidal *Posidonia* seagrass meadows. The high standing stock of e.g., *Cibicides lobatulus*, *Lituotuba lituiformis* and *Planorbulina mediterranensis* of the Kosterfjord experiment is not reported in subrecent assemblages of surface to subsurface sediment samples from other fjord settings.

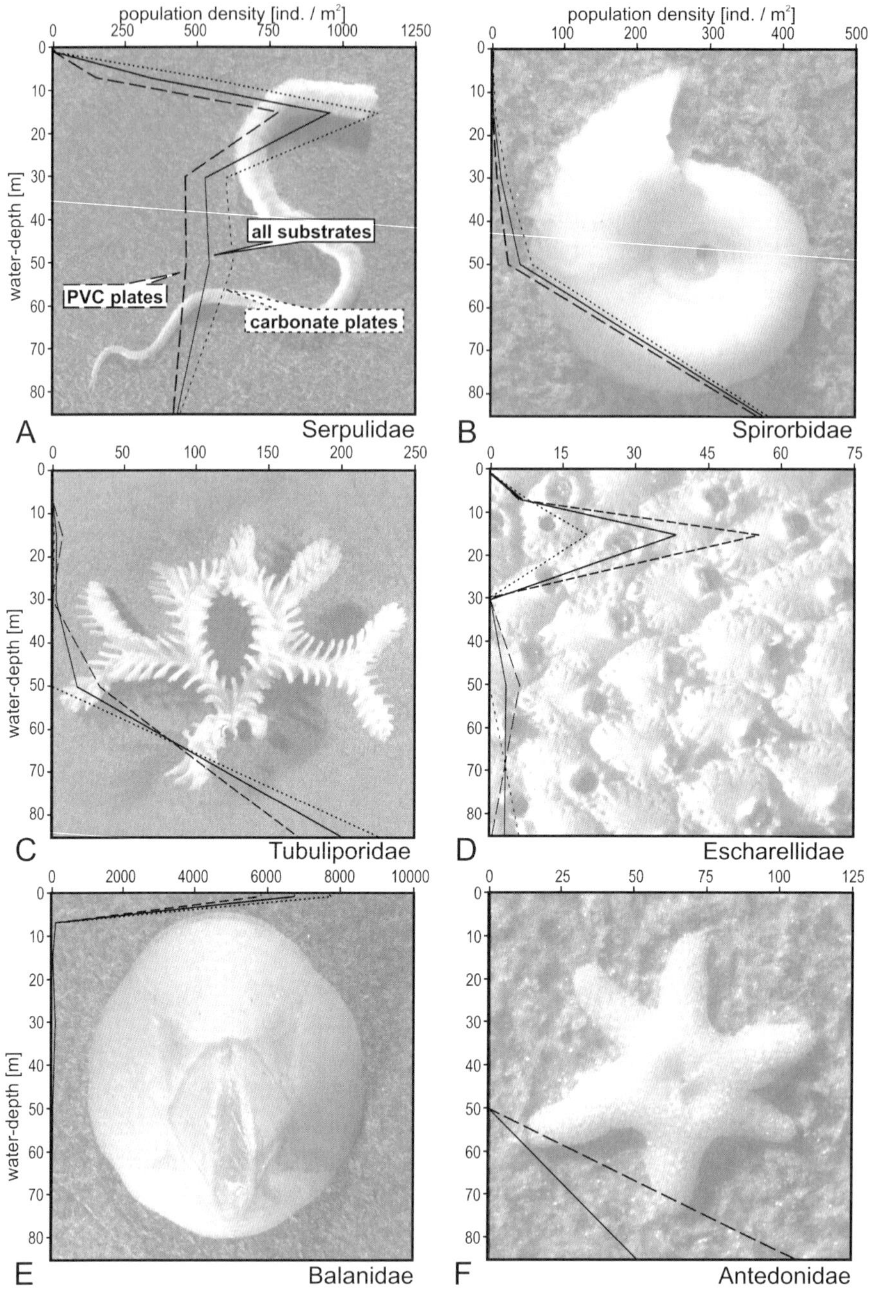

Fig. 45 Arithmetic mean of population densities of all exposure times on the different substrate types for the Serpulidae (**A**), Spirorbidae (**B**), Tubuliporidae (**C**), Escharellidae (**D**), Balanidae (**E**) and Antedonidae (**F**)

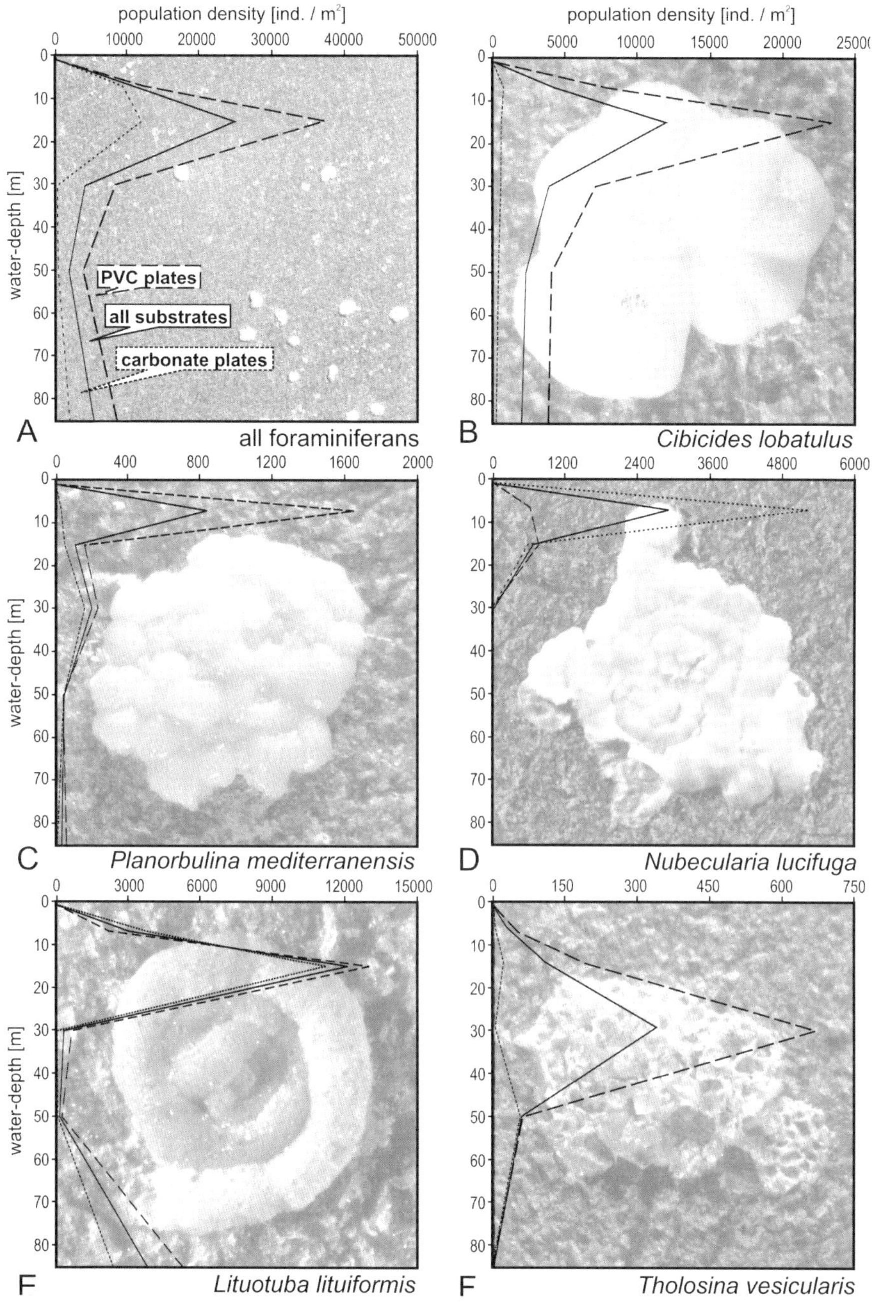

Fig. 46 Mean population densities of all exposure times on the different substrate types for all foraminiferans (**A**), *Cibicides lobatulus* (**B**), *Planorbulina mediterranensis* (**C**), *Nubecularia lucifuga* (**D**), *Lituotuba lituiformis* (**E**) and *Tholosina vesicularis* (**F**)

This is because hard substrates are known to support assemblages different from soft bottom ones, characterised by the presence of attached and clinging species. The dilution of epiphytic species in assemblages of fjord sediments may be the result of naturally limited availability of firm substrates, which are rarely sampled in field studies. Furthermore, the analysis of surface sediments from the nearby Koljö Fjord between 12 m and 43 m water depth, where *Elphidium* species dominate the assemblages at all water depths (Gustafsson & Nordberg 1999), shows no indication for any contribution of the five dominant species of our study to the fossil assemblage. Conversely, *Elphidium* species are very rare on the settlement panels in the Kosterfjord although reported to follow an epibenthic mode of life (Wefer & Lutze 1978). However, more recent studies observed a wide range of epi- to infaunal microhabitats for different *Elphidium* species in soft and fine-grained sediments of brackish coastal environments rather than on hard substrates (e.g., Corliss & Van Weering 1993; Hunt & Corliss 1993; Linke & Lutze 1993; Wollenburg 1995), which explains the very rare occurrence of *Elphidium* on our experimental plates.

Langer (1993) describes a similar assemblage of *Cibicides lobatulus*, *Planorbulina mediterranensis*, *Rosalina* spp. and *Nubecularia lucifuga* from elevated substrates (leaves and blades of *Posidonia*, *Sargassum*, *Ectocarpus*, etc.) around Volcano Island in the Mediterranean Sea at 5-40 m water depth. He differentiated epiphytic foraminiferans into different morphotypes according to test shape and their ecological preferences. The distribution of permanently sessile foraminiferans (*Planorbulina mediterranensis*, *Nubecularia lucifuga*), belonging to 'Morphotype A', seem to be related to plants with comparatively long life-spans and large, flat or arch-shaped surfaces of the leaves. The depths distribution of these species depends on the occurrence of the plants between 15 m and 30 m. In our study, the maximum abundance of *Planorbulina mediterranensis* and *Nubecularia lucifuga* is quite similar between 7 m and 30 m. Dominant agglutinated species are *Lituotuba lituiformis* and *Tholosina vesicularis*, which can be attributed to 'Morphotype A' being permanently sessile. A group of temporary attached foraminiferans are described as 'Morphotype B' with *Rosalina* spp. and *Cibicides lobatulus* being the most abundant species (Langer 1993). Their depths range is similar (5-35 m), but these species are temporary motile providing the ability to react actively on environmental changes. In contrast to Langer (1993), Gross (2000) observed in microcosm experiments that *Planorbulina mediterranensis* is not permanently sessile. This species rested within cysts for periods of weeks or months, after which the cysts were abandoned and new ones constructed. The distribution of

Cibicides lobatulus in the Kosterfjord experiment shows a maximum at 15 m after 24 months, but is also very abundant between 15 m and 50 m.

Substrate preference

The higher population densities on the PVC plates versus the carbonate plates reflect a significant substrate preference. In particular, *Cibicides lobatulus* was found to be 25 times more abundant on the PVC plates. This preference seems counterintuitive, since one could expect a reverse distribution pattern on the carbonate substrate due to the allowance for a better fixation in depressed attachment scars. *Planorbulina mediterranensis* and *Tholosina vesicularis* apparently also prefer PVC whereas for *Lituotuba lituiformis* is about equally distributed on both substrate types and *Nubecularia lucifuga* favours the calcareous substrate. These substrate preferences either owe to different attachment and/or reproduction strategies or do reflect possible interactions with bioeroding organisms weakening the fixation on the carbonate plates.

Foraminiferans as bioeroding agents

A number of foraminiferans are known from the fossil record and Recent environments to settle on hard substrates, leaving specific etched attachment scars, as for instance *Rosalina globularis* (Delaca & Lipps 1972; Mullineaux & Delaca 1984), *Cibicides refulgens* (Mullineaux & Delaca 1984; Alexander & Delaca 1987), *Planorbulinopsis parasita* (Banner 1971) and *Hyrrokkin sarcophaga* (Cedhagen 1994; Freiwald & Schönfeld 1996). A compilation on bioeroding foraminiferans is given by Vénec-Peyré (1996). A large variety of organic and inorganic substrates have been found infested by boring foraminiferans such as wood, mollusc shells, bryozoan skeletons, crustacean carapaces, corals and even other foraminiferal tests (Vénec-Peyré 1996). The characteristic etching trace produced by the very rare *Gypsina vesicularis* as recorded during the course of the Kosterfjord experiment has not been reported before.

The attachment scars represent by far the most abundant macroboring trace with more than 80,000 scars per square metre recorded at 15 m water depth after two years exposure (see also Wisshak et al. 2005b). Hence, foraminiferans contribute to a considerable portion to the initial stages of bioerosion in this and potentially also in other cold-temperate settings. During advanced stages of bioerosive hard substrate degradation, however, other bioerosion agents (predominantly boring sponges) clearly outcompete foraminiferans as bioeroders.

Environmental controls

The absence of any foraminiferans at the shallowest station (1 m) can be attributed to a unsuitable hydrodynamic regime, unsuitable profound temperature and salinity variations (~0-20°C; ~8-31 psu), and/or to the strong spatial competition by balanids, which densely occupied both substrate types already after 12 months exposure. At the 7 m station, salinities frequently drop to values below 25 psu, also coupled with strong temperature fluctuations (~1-18°C). Consequently, the foraminiferans thriving well at this depth (*Cibicides lobatulus*, *Planorbulina mediterranensis* and *Nubecularia lucifuga*) can be regarded as considerably euryhaline and eurytherm. The spatial competition is less pronounced than at 1 m water depth but is still considerable owing to dense epilithic algae cover. At 15 m water depth, at the population density maximum, the hydrologic variations are progressively less pronounced and the spatial competition with other epibiota is decreasing. This leads to very favourable conditions especially for *Cibicides lobatulus* and *Lituotuba lituiformis*. Below the thermocline and halocline separating the Baltic outflow and the Atlantic inflow waters, open marine conditions prevail with much smaller fluctuations of temperature and salinity. Here, increased species diversity is promoted along with lower specimen densities. A slight increase in specimen numbers at the 85 m station (especially of *Lituotuba lituiformis*) can be attributed to the proximity to the Säcken Reef with its enhanced biodiversity and/or to the stronger current regime on the distinct sill (max. 24 cm/s; mean 4.5-6.0 cm/s).

Tholosina vesicularis and especially *Cibicides lobatulus* are the first and exclusive species to settle on the panels after 6 months exposure. Thus they can be classified as r-strategists (opportunists). A more diverse foraminiferal assemblage is not developed until 12 months and the highest diversity is recorded after 24 months exposure. The overall specimen numbers are increasing in parallel with time. This time scale resembles recovery periods recorded in soft-bottom settings. The experiments by Wefer & Richter (1976) found ambient densities 13 months after deployment in the Eckernförder Bay (Western Baltic Sea). Alve (1999) recorded a minimum of 8 months in the Oslofjord (S Norway). She attributed the comparably long recovery periods combined with the development of opportunistic assemblages (similar to metazoan pathway) to low current regimes (<10 cm/s; Alve 1999). In the case of hard substrates, however, this pattern may be considered as a typical strategy, since dispersal through release of embryonic juveniles represents the main reproductive strategy as opposed to passive suspension and transport of various soft bottom foraminiferal growth stages (Alve 1999).

6 Quantitative bioerosion and carbonate accretion

6.1 Assessing bioerosion and carbonate accretion rates

A centrepiece of the Kosterfjord experiment is a quantitative analysis and evaluation of the two opposing processes bioerosion and carbonate accretion. Both rates yield an important integral for the interpretation of the carbonate cycling in this cold-temperate setting.

For the determination of the bioerosion and carbonate accretion rates a gravimetrical approach was applied and the values are expressed in eroded or accreted grammes $CaCO_3$ per square metre and year ($g/m^2/y$). In the following tables the determined rates at each depth and exposure time are given with their means, standard deviation and number of retrieved plates.

Bioerosion rates

The bioerosion rates (Fig. 47; Table 10) range from a mean of zero (85 m after 24 months) to a maximum of -218 $g/m^2/y$ (lagoon after 6 months).

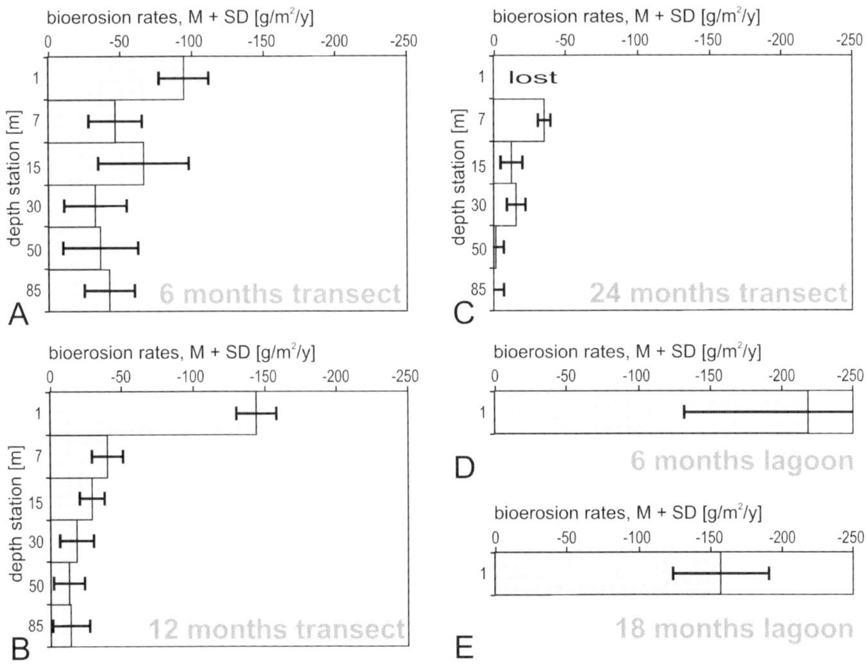

Fig. 47 Graphic display of the mean (M) bioerosion rates per depth and exposure time with the standard deviation (SD) at the transect after 6 months (**A**), 12 months (**B**) and 24 months (**C**) exposure as well as for the lagoonal setting after 6 months (**D**) and 18 months (**E**) exposure

Table 10 Bioerosion and carbonate accretion rates at each depth and exposure time given as mean values (M) with standard deviation (SD) and number of retrieved plates (n). In addition, the total mean per exposure time, and the overall total means of all exposure times and depths is given (lagoon stations excluded)

Water depth [m]	Exposure time [days]	Total bioerosion [g/m²/y]	Total carbonate accretion [g/m²/y]
1 L	178	-218 ±86 (n=6)	1.02 ± 3.03 (n=12)
1 L	513	-157 ±34 (n=5)	0.48 ± 0.97 (n=11)
1	177	-94 ±17 (n=5)	0.00 ± 0.00 (n=11)
1	389	-144 ±14 (n=6)	361.93 ±67.35 (n=12)
1	lost	(n=0)	(n= 0)
7	178	-47 ±19 (n=6)	0.14 ± 0.10 (n=12)
7	392	-40 ±11 (n=6)	0.10 ± 0.03 (n=12)
7	726	-36 ± 5 (n=5)	0.86 ± 1.04 (n=11)
15	178	-66 ±31 (n=6)	1.01 ± 0.49 (n=12)
15	392	-29 ± 9 (n=6)	0.74 ± 1.28 (n=12)
15	726	-13 ± 8 (n=4)	1.29 ± 1.25 (n=10)
30	175	-32 ±22 (n=6)	0.00 ± 0.00 (n=12)
30	390	-18 ±12 (n=6)	0.13 ± 0.06 (n=12)
30	724	-17 ± 6 (n=6)	0.14 ± 0.23 (n=12)
50	180	-36 ±26 (n=6)	0.05 ± 0.09 (n=12)
50	390	-13 ±11 (n=6)	0.10 ± 0.03 (n=12)
50	724	-2 ± 5 (n=6)	0.07 ± 0.03 (n=12)
85	177	-43 ±17 (n=6)	0.09 ± 0.10 (n=12)
85	390	-14 ±13 (n=6)	0.73 ± 0.83 (n=12)
85	725	0 ± 8 (n=6)	0.11 ± 0.07 (n=12)
1-85	~6 months	-54 total	0.22 total
1-85	~12 months	-43 total	60.63 total
1-85	~24 months	-13 total	0.49 total
1-85	6-24 months	-37 overall total	19.42 overall total

The highest bioerosion rates are found at the 1 m stations and especially in the lagoonal setting. With one exception (7 m after 6 months), the bioerosion rates at all deeper stations did not exceed -50 g/m²/y. At all depths, the rates were decreasing with time except for 1 m from 6 to 12 months exposure. This time-dependent decrease is also evident for the total means per exposure time from -54 g/m²/y (6 months) to -43 g/m²/y (12 months), and finally to -35 g/m²/y (24 months) when assuming constant rates at the lost 1 m station, and -13 g/m²/y when excluding this depth from the mean calculation. The overall total mean of all exposure times and depths is -37 g/m²/y. The standard deviations are generally moderate to high, often more than 50% of the mean value but rarely exceeding the latter (only for the lowest two rates recorded).

Carbonate accretion rates

The accretion rates (Fig. 48; Tables 10-11) range about two magnitudes lower from zero to 1.29 g/m²/y. One extreme exception is represented by the 1 m station after 12 months, where massive encrustation by balanids (see below) accounts for 361.93 g/m²/y. Even when excluding this depth from the calculation, there is an overall increase in mean accretion values with time from 0.26 g/m²/y (6 months) to 0.36 g/m²/y (12 months) to 0.49 g/m²/y (24 months). However, this increase is not systematic for all depths. The overall total mean of all exposure times and depths sums up to 19 g/m²/y. The standard deviation is generally high, often exceeding the mean values.

The limestone plates yield slightly higher overall accretion rates (20.99 g/m²/y) than the PVC plates (17.89 g/m²/y). This trend holds true for all exposure times and most of the individual frames except for a few of the deeper stations after 12 and 24 months exposure.

At all depths and exposure times (except for 1 m after 12 months) a net bioerosion (accretion rate minus bioerosion rate) is found.

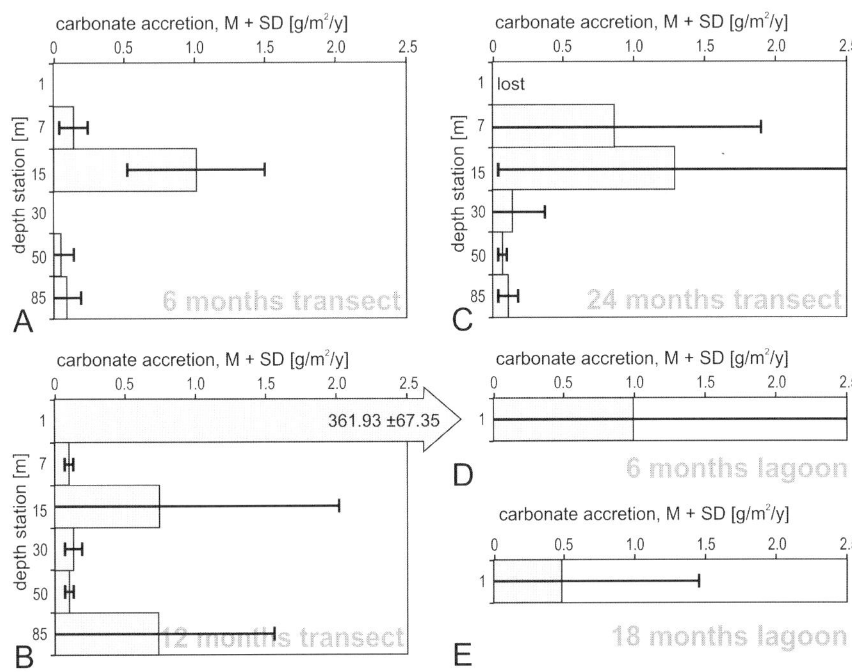

Fig. 48 Graphic display of the mean (M) carbonate accretion rates per depth and exposure time with the standard deviation (SD) at the transect after 6 months (**A**), 12 months (**B**) and 24 months (**C**) exposure as well as for the lagoonal setting after 6 months (**D**) and 18 months (**E**) exposure

Table 11 Carbonate accretion rates on limestone and PVC plates at each depth and exposure time given as mean values (M) with standard deviation (SD) and number of retrieved plates (n). In addition, the total mean per exposure time, and the overall total means of all exposure times and depths is given (lagoon stations excluded)

Water depth [m]	Exposure time [days]	Carbonate accretion limestone plates [g/m²/y]	Carbonate accretion PVC plates [g/m²/y]
1 L	178	1.84 ± 4.11 (n=6)	0.20 ± 0.37 (n=6)
1 L	513	0.78 ± 1.26 (n=5)	0.23 ± 0.52 (n=6)
1	177	0.00 ± 0.00 (n=5)	0.00 ± 0.00 (n=6)
1	389	388.51 ±69.97 (n=6)	335.35 ±52.57 (n=6)
1	lost	(n=0)	(n=0)
7	178	0.14 ± 0.10 (n=6)	0.14 ± 0.10 (n=6)
7	392	0.11 ± 0.04 (n=6)	0.09 ± 0.00 (n=6)
7	726	0.87 ± 0.86 (n=5)	0.86 ± 1.16 (n=6)
15	178	1.27 ± 0.53 (n=6)	0.75 ± 0.25 (n=6)
15	392	1.19 ± 1.69 (n=6)	0.29 ± 0.11 (n=6)
15	726	1.54 ± 1.60 (n=4)	1.12 ± 0.91 (n=6)
30	175	0.00 ± 0.00 (n=6)	0.00 ± 0.00 (n=6)
30	390	0.11 ± 0.04 (n=6)	0.16 ± 0.07 (n=6)
30	724	0.06 ± 0.05 (n=6)	0.23 ± 0.30 (n=6)
50	180	0.07 ± 0.10 (n=6)	0.03 ± 0.08 (n=6)
50	390	0.11 ± 0.04 (n=6)	0.09 ± 0.00 (n=6)
50	724	0.06 ± 0.02 (n=6)	0.07 ± 0.04 (n=6)
85	177	0.18 ± 0.08 (n=6)	0.00 ± 0.00 (n=6)
85	390	1.25 ± 0.91 (n=6)	0.22 ± 0.09 (n=6)
85	725	0.08 ± 0.04 (n=6)	0.14 ± 0.08 (n=6)
1-85	~6 months	0.28 total	0.15 total
1-85	~12 months	65.22 total	56.03 total
1-85	~24 months	0.63 total	0.48 total
1-85	6-24 months	20.99 overall total	17.89 overall total

Foraminiferal carbonate accretion rates

In addition to the gravimetrical determination of non-specific carbonate accretion rates, an estimation of carbonate accretion rates by *Cibicides lobatulus*, representing the most abundant calcareous foraminiferan species, was undertaken based on the average test weight and species distribution. The test sizes range between 100 µm and 1,800 µm in diameter. The average test weight is 0.0187 mg (n = 250) and the standing stock varies between 0 and 17,500 ind./m². The estimated carbonate production of *Cibicides lobatulus* varies between 0 and 0.326 g/m²/y. After 6 months exposure, the carbonate production of *Cibicides lobatulus* on the PVC plates is confined to

15 m and below, and on the carbonate plates to the 85 m station. The values are relatively low and the maximum was recorded at 15 m water depth with 0.021 $g/m^2/y$ (Table 12).

Table 12 Carbonate accretion rates in grammes per square metre and year ($g/m^2/y$) of *Cibicides lobatulus* after 6, 12, and 24 months on PVC and carbonate plates

Depth	PVC plates			Carbonate plates		
[m]	6 months	12 months	24 months	6 months	12 months	24 months
1	0	0	-	0	0	-
7	0	0.086	0.086	0	0.002	0.009
15	0.021	0.326	0.234	0	0.010	0.004
30	0	0.321	0.007	0	0.007	0.004
50	0.008	0.057	0.040	0	0.005	0.003
85	0.007	0.045	0.039	0.001	0.003	0.001

A temporal increase in carbonate accretion is expressed during the first 12 months on all settlement plates. In the second year of exposure, the rates were declining at all depths except for the 7 m station. A general spatial increase in carbonate accretion occurs from shallow depths to mid-water depths (7 to 30 m) but values decline again towards the deeper stations. The comparison between carbonate accretion rates on PVC and carbonate plates indicates rates up to two magnitudes higher on the PVC plates, corresponding to the pronounced substrate preference (see above).

6.2 Bioerosion rates discussion

Methodology

The comparison of the Kosterfjord bioerosion rates with those reported in the literature is complicated by the varying methodological approaches applied in previous quantitative bioerosion studies. They vary from different image-analysis techniques to gravimetrical quantification methods and are far from being standardised. As a consequence, bioerosion rates are either referring to volume or surface area removed (absolute or per time interval) and were consequently given in many different units. In addition, different substrate mineralogies, densities and sizes were applied ranging from micrite blocks to coral or molluscan carbonate (with varying degrees of organic compounds) in sand size up to decimetre large test blocks.

In contrast to previous studies, true 2D test surfaces are applied in the Kosterfjord experiment by sealing off all but the upper side of the micrite blocks with epoxy resin. By employing a gravimetrical quantification method, it was possible to directly obtain weight values removed per surface and exposure time, expressed as grammes per square metre and year ($g/m^2/y$), without the need of interpolating from three dimensional test blocks. By using 5 to 6 test blocks at each depth with an overall relatively large surface area of about 5 to 6 dm^2, the statistical accuracy of the obtained values was further enhanced.

The widely used method to quantify bioerosion and accretion with the aid of image analysis of sectioned 3D test blocks (as introduced by Kiene 1985) requires interpolation of the surface area and calculated weight losses via substrate densities and calculated removed volume, respectively. In addition, complications arise from a decrease in available surface with time and interference of borings entering the test blocks from different sides, requiring further interpolations (which have rarely if at all been made). In summary, this procedure requires too many interpolations and/or indirect measurements, and consequently governed the choice to apply the modified direct gravimetrical approach as outlined above. A direct comparison of results by the classical image-analysis method with the refined gravimetrical approach would be an appreciable task for future studies.

A drawback of the gravimetrical method is, that it bears some inaccuracy for early stages of microborer colonisation, where only few milligrammes per test bock are removed, pushing the results towards the error range of the method due to a maximum accuracy of the precision scale ±0.01 g for samples >100 g and further potential errors due to the H_2O_2-preparation and handling of the plates during deployment and recovery of the panels. The comparatively high relative standard deviations for the lowest rates obtained can be partly attributed to this effect. However, bathymetric and temporal patterns of bioerosion rates are clearly expressed and backed up by the qualitative and semi-quantitative data obtained by SEM image analysis.

Microbioerosion versus macrobioerosion versus grazing rates

Further attention has to be drawn to the methodological complication in separating bioerosion intensities of microborers, macroborers and grazers. When applying image analysis, one has to bear in mind, that grazing removes substrate that was already bored by micro- as well as macroborers and thus bioerosion rates of the latter are systematically underestimated while grazing activity is overestimated. This should always be taken into consideration when interpreting and comparing such rates from the literature.

When applying a gravimetrical approach as it has been done in the case of the Kosterfjord experiment, only total bioerosion and carbonate accretion are measured. However, the qualitative analysis of the contributing micro-

and macroborers demonstrated, that traces of grazers are rare and the degree of grazing is very low if compared to tropical environments due to relative scarcity of grazing echinoids and herbivorous fish. Only chitons and/or gastropods played a minor role as recorded in few erosion plates exhibiting radula rasping marks. Also, grazing is limited primarily to the euphotic zone triggered by the need of epilithic algal cover for the grazers to feed on. It has furthermore been shown that macroborers play a minor role except for the shallowest stations (1 and 7 m), where polychaete borings were abundant and most likely controlled the bioerosion rates. Only exception of this observation are the foraminiferal attachment scars found at all depths. Hence, the assessed bioerosion rates can be considered as an approximation for microbioerosion rates, except for the shallowest stations (1 and 7 m) where they represent a composite value of micro- and macrobioerosion with a low contribution by grazers.

Substrate composition

Concerning the influence of the experimental substrate mineralogy and density, contradicting results are reflected in the literature: Vogel et al. (1996, 2000) found the highest degree of microbioerosion in the micritic substrates when compared to molluscan (*Strombus*) substrates (520 $g/m^2/y$ versus 90 $g/m^2/y$). They interpret this discrepancy by the organic layers in the molluscan shell ultrastructure possibly hindering free growth of phototrophs but favouring the distribution of heterotrophs. Following this line of reasoning, we would expect overestimated bioerosion rates for the Kosterfjord experiment in the photic and underestimated rates in the dysphotic and aphotic zones when compared to organic-bearing calcareous substrates.

In all other studies considering different substrates chiefly macroboring and grazing activity was investigated. When comparing micrite substrates with those from *Porites* blocks, Reaka-Kudla et al. (1996) found significantly higher degree of total bioerosion of the *Porites* test blocks (mean total erosion at 5 to 13 m water depth of 25,400 $g/m^2/y$ versus 4,100 $g/m^2/y$ in micrite substrates) during their 15 months experimental study on the Galápagos Islands. Scott et al. (1988) found a higher degree of bioerosion in Pleistocene coral rubble blocks than in concrete blocks of 4.5% compared to 3% by volume after 13 years exposure in a shallow tropical lagoon (Jamaica). Neumann (1966) found no differences in the boring rate between different substrate mineralogies but between substrate densities in his experiments on boring sponges with a higher degree of bioerosion in denser coral skeletons. This finding was confirmed by Highsmith (1981) for boring bivalves but contradicted by Hutchings (1981) in the case of polychaetes. Rützler (1975) compared gravimetrically determined sponge bioerosion rates of Iceland spar and molluscan shells, concluding that porosity of the substrate aids quicker and deeper penetration but no more material to be removed per time unit. Whether this finding is generally valid for other agents of bioerosion is doubtful, considering the variety of different modes of penetration. In summary, it can be expected that substrate composition and density has differing influences on different agents of bioerosion and generalisations should be avoided.

Substrate orientation

To date very little is known on the influence of the substrate orientation on the rate of bioerosion. Kiene (1988) and Hassan (1998) have shown, that grazing (in both cases dominated by herbivorous fish) is more pronounced on the upper side whereas the lower side was primarily grazed by smaller invertebrates (e.g., gastropods and chitons). There are no observations available as yet revealing the difference in macro- and microendoliths with respect to substrate orientation. Their distribution is primarily governed by the light availability, so one could expect higher rates on the upward facing side of the panels in the photic zone. However, the upward facing side is prone to sedimentation and grazing, both in turn negatively influencing larval recruitment. To understand this interrelationship, further experimental studies are to be undertaken. In the present study all bioerosion rates refer solely to the upward facing side (in contrast to most previous studies), which is grounded in the aim to correlate bioerosion rates to the measured light availability.

Substrate size

In the case of major substrate size differences, one can expect differences in the rate of bioerosion as for instance when comparing the sand size grains applied by Tudhope & Risk (1985) with the centimetre- to decimetre-sized *Porites* or micrite blocks applied in most other studies. It can be assumed that a varying potential in macroborer recruitment and accessibility to grazers accounts for this influence. Hutchings (1985) and Kiene & Hutchings (1994b) found no differences between substrates of moderately (max. twofold) varying size in their experiments. Hence, in this respect, the substrate size (10 x 10 x 1 cm) employed in the present study can readily be compared with most previous studies.

General precaution

In summary, comparisons of absolute bioerosion and carbonate accretion rates may be drawn between studies as long as they are not overstressed and when taking the methodological pitfalls outlined above into account. Nevertheless, the validity and value of relative trends and patterns observed within one study site – which can in turn be interpreted in relation to other study sites – is unbroken.

However, it must be stressed, that experimentally determined bioerosion rates in general are only an approximation for a specific substrate, water depth and exposure time and balancing over whole reefs or other carbonate sedimentary settings should be undertaken with caution. It can be regarded as a general pattern that bioerosion values may vary significantly between study sites, settings and even experimental frames due to patchy dispersal and recruitment of bioerosion agents (e.g., Kiene & Hutchings 1994a).

The published bioerosion rates are summarised in Table 1 (Sect. 1.3) together with the methodology applied and substrates used. Where necessary and possible, the values were translated into $g/m^2/y$.

Bioerosion rates and bathymetry

In his review on the factors controlling bioerosion, Scott (1988: 684) stated "The evidence is fairly conclusive that there is no relationship between depth and bioerosion". This conclusion, however, was based only on studies in shallow-water tropical reef settings (e.g., Bromley 1978; Highsmith 1981 and) and is at least in its generalisation certainly not true. It merely only applies to the uppermost few metres of the water column, where different reef environments are prone to, e.g., significantly different hydrodynamic regimes, tidal influences and variations in substrate species distribution and availability, here primarily governing the rate of bioerosion. The variability of bioerosion rates between reef flat, lagoonal patch reefs, leeward and windward shallow slopes etc. has been demonstrated in numerous experimental studies (e.g., Kiene & Hutchings 1994a and b; Vogel et al. 2000; Pari et al. 1998, 2002).

In the present setting, we see a significant decrease in bioerosion rates with depth, which can be related with confidence to the light availability controlling the distribution of phototrophic microendoliths and the distribution of grazers feeding on them. This relationship was also clearly expressed in the microbioerosion rates determined in the Bahamas experiments carried out by Vogel et al. (2000) along a transect from the intertidal down to 275 m water depth and by Hoskin et al. (1986) in a transect from the intertidal down to 477 m water depth. In addition, the relationship of light availability and microendolithic distribution and abundance is well expressed in the qualitative analysis and has been shown in numerous previous studies (e.g., Golubic et al. 1975; Zeff & Perkins 1979; Budd & Perkins 1980; Glaub 1994).

The same pattern applies to subtidal macroborer and grazing activity as illustrated for instance by Kobluk & Kozelj (1985) who found a clear correlation of decreasing boring intensities of macroborers with increasing depth along a transect down to 34 m at Bonaire (Netherland Antilles) or Hassan (1998) who saw an increase in bioerosion between 5 and 15 m but a significant decrease between 15 and 40 m water depth in her Red Sea experimental study. In the Kosterfjord experiment, most macroborer traces are scarcely distributed and consequently no clear bathymetrical trend can be recognised. Nevertheless, the distribution of polychaete borings (*Caulostrepsis* isp.) clearly decreases with depth as does the abundance of foraminiferal attachment scars below their maximum at 15 to 30 m water depth.

A number of authors observed in qualitative investigations, that bioerosion of skeletal carbonate in high-latitudes is apparently higher in calm and deeper water bodies compared to agitated shallow-water settings (e.g., Alexandersson 1972; Scoffin et al. 1980a). On the first sight, this observation contradicts the decrease in bioerosion rates with depth recorded in the present and many previous quantitative and qualitative studies. But, in the case of the experiments, the substrates are fixed and not free moving and they are slightly elevated and thus less prone to background sedimentation than the surrounding sea floor.

Bioerosion rates and nutrient supply

The dependence of bioerosion rates and eutrophication is a complex interplay with different agents of bioerosion being inversely affected. The contradicting results of previous studies (see below) reflect the complexity of this interrelationship and the difficulty to separate eutrophication from other controlling factors.

It has been shown, that bioerosion rates and diversity is significantly enhanced by eutrophication (see Hallock 1988 for a review on this subject). Highsmith (1980) saw a positive global correlation of plankton primary productivity and the number of boring bivalves and suggested this pattern to apply for all planktivorous borers. The positive correlation of bioerosion rates and eutrophication was partly confirmed and partly contradicted by the long-term (5 years) experiment in French Polynesia (Le Bris et al. 1998;

Pari et al. 2002), where epilithic algae cover and grazing was enhanced in high-nutrient areas whereas endolithic algae were more abundant in oligotrophic sites. In contrast, in their experimental study at La Réunion, Chazottes (1996) and Chazottes et al. (2002) observed an inverse complex interplay of eutrophication and various bioerosion agents governed by the epilithic algae that influences grazing activity, which in turn controls recruitment of macro- and microborers. They observed the lowest rates of grazing and macrobioerosion but the highest rates of microbioerosion in nutrient-rich areas associated with high macroalgae cover. Yet another study found no correlation – the ENCORE experiment at the Great Barrier Reef (Kiene et al. 1995; Kiene 1997; Vogel et al. 2000) – which yielded only minute and insignificant impact of nutrient enhancement (nitrogen and phosphorous) on bioerosion rates and microendolithic community structures. They consequently suggested that the (individual) micro-milieu within the microendolithic cavities buffers nutrient influx. It can be speculated that this effect does not apply to macroborers, resulting in their nutrient dependency.

In any way, the dependence of bioerosion rates and eutrophication only applies at least for comparable biogeographic ranges, prohibiting a comparison of the literature data with the Kosterfjord setting. On a global scale other factors such as light availability and water temperatures as well as significantly different biocoenoses most likely outperform nutrient levels as controlling factors. Consequently, the results of the Kosterfjord experiment may only be compared to similar temperate sites with differing nutrient fluxes but no such experimental data is available to date.

Bioerosion rates and exposure time

The results of the Kosterfjord experiment show a general decrease in bioerosion rate with time at all depth stations (except for 1 m between 6 and 12 months exposure). This can be attributed to a decrease in bioerosion rate of microendoliths after initial infestation of the pristine substrate and also after establishment of a mature ichnocoenoses (in the case of the shallow stations). An additional factor can be the successive shading of phototrophic endoliths by epilithic algae cover in the photic zone. This result corresponds to the findings of Vogel et al. (1996, 2000) who also saw a decrease in microbioerosion rate between 6 months and 24 months exposure after an initial increase between 3 and 6 months. Only below 100 m water depth they found slightly increasing bioerosion rates over the course of the 24 months experimental period.

The total bioerosion rates obtained by most other previous studies increased with time (e.g., Peyrot-Clausade et al. 1995b; Kiene 1985, 1988; Kiene & Hutchings 1994b; Hassan 1998; see Table 1), which is due to the absolute dominance of grazing and to a lesser degree of macrobioerosion in those settings. The only study where microbioerosion rates were separately determined – the initial 2-year experimental study at Moorea (French Polynesia) by Peyrot-Clausade et al. (1995a) – showed a decrease with time in accordance to the present results.

It needs to be taken in mind, that temporal trends in bioerosion rates during relative short-time deployments are influenced by seasonal recruitment patterns of plants and invertebrate larvae. The time a substrate becomes available consequently influences the composition of the boring biocoenoses and inevitably also the rate of bioerosion (Hutchings 1981, 1985; Hutchings & Bamber 1985). The present study site for instance is subjected to seasonal algae blooms, which likely affect the recruitment of endolithic species.

Tropical versus cold-temperate bioerosion rates

As a general result of the quantification, we clearly see a significantly lower pace and rate of bioerosion in the cold-temperate Kosterfjord site if compared with tropical reef settings. The determined bioerosion rates are to a high degree governed by microbioerosion except for the shallowest stations (1 and 7 m), where macroborers (chiefly polychaetes) are additionally relevant. The results are thus best comparable to the results of the Bahamas experiments conducted by Hoskin et al. (1986) and Vogel et al. (1996, 2000) which both also applied micritic substrates, determined specific microbioerosion rates and studied an extended bathymetric transect (Table 1). Hoskin et al. (1986), who also applied a gravimetrical quantification method, found microbioerosion rates ranging from -259 g/m^2/y in the shallow water to virtually zero at the deepest waters (477 m water depth). Vogel et al. (1996, 2000) applied image analysis for quantification purposes and recorded even higher rates of microbioerosion with a maximum mean value of -520 g/m^2/y at the shallowest station after 6 months exposure and a decrease down to negligible values at the deepest station (275 m water depth). Hence, the degree of microbioerosion at the tropical Bahamas site can be judged as roughly two to ten times as high if compared to the Kosterfjord setting, where most rates do not exceed -50 g/m^2/y.

From the GBR (Australia), two very differing results concerning microbioerosion activity were reported in the literature: Kiene (1997) found only -8 to -43 g/m^2/y while Tribollet et al. (2002) more recently reported -120 to -1,340 g/m^2/y, both applying image analysis on *Tridacna* and *Porites* substrates, respectively, deployed in shallow waters (<10 m). Peyrot-Clausade et al. (1995a) reported mean values comparable to the Bahamas results of -140 to -570 g/m²/y from a barrier reef flat fringing Moorea (French Polynesian). From the Indian Ocean, Chazottes (1996) reported mean microbioerosion rates of -24 to -69 g/m²/y. Even though all these studies applied image-analysis for quantification purposes (which is tentatively underestimating microbioerosion rates) and used *Porites* or *Tridacna* substrates, they indicate an at least comparable but probably generally higher degree of microborer activity in these tropical reef settings.

As far as overall bioerosion rates go, the tropical rates are generally far exceeding the Kosterfjord rates, since in the Tropics, macroborer and especially grazers chiefly govern the degree of bioerosion. As a consequence, the total bioerosion rates often exceed 1,000 g/m²/y (Table 1) and the highest overall bioerosion rate reported in the literature amounts to an impressive -25,400 g/m²/y (Galápagos fringing reef; Reaka-Kudla et al. 1996). Hence, overall bioerosion rates in the tropical reef settings can be judged to exceed the cold-temperate Kosterfjord values by one to two orders in magnitude.

6.3 Carbonate accretion rates discussion

Substrate composition

The substrate type is a controlling factor on the intensity of encrustation, both in terms of material used and surface roughness. Harriott & Fisk (1987), as an example, found significant differences in recruitment (number and taxa) of scleractinian corals on eight different substrate types. Concerning the substrate roughness, Gunkel (1997) demonstrated considerable differences in invertebrate recruitment on smooth versus rough PVC plates and limestone plates. While the recruitment diversity was highest on the limestone plates and lowest on the smooth PVC substrates, the specimen numbers of the various epizoans were affected differentially. This is owing to specific larvae behaviour of the various encrusting invertebrate groups and taxa – a topic extensively reviewed by Crisp (1974).

In the present study, different carbonate accretion rates on the limestone plates compared to the PVC plates were recorded, accordingly. In Sect. 5.3, it has been shown that the various calcareous epizoans show a differential substrate preference, expressed in varying population densities on both substrate types. Even though the majority of the point-counted epizoan groups favours PVC or has no significant preference, the quantified overall carbonate accretion rates are slightly higher on the carbonate substrates. This is owing to the fact, that for instance the serpulimorphs and balanids with relatively heavy individual skeletons (especially if compared to the various foraminiferal taxa) prefer the carbonate substrate and thus steer the overall accretion rates.

Hard- versus soft-bottom productivity of foraminiferans

In the Kosterfjord, the carbonate accretion of *Cibicides lobatulus* varies between 0 and 0.326 $g/m^2/y$ (Table 12) and the maximum occurs within the transitional layer between the Baltic outflow and the Atlantic inflow water mass at 15 to 30 m water depth. The maximum rates are around two magnitudes higher than in the western Baltic Sea, where Wefer & Lutze (1978) reported 0.002 to 0.0035 $g/m^2/y$ for the epiphytic foraminiferans *Ophthalmina kilianensis*, *Elphidium excavatum* and *Elphidium gerthi*. Epiphytic species only occurred in the turbulent zone (<12 m) on the erosional platform closely related to algae. Wefer & Lutze (1978) included the probable reproduction rate per year into their calculation, which ranges between 1 and 1.3 depending on the species. However, in the Kosterfjord calculations the reproduction rate of *Cibicides lobatulus* is not applicable, since it was not differentiated between living and dead individuals on the date of recovery. Langer (1993) attributes species of 'Morphotype B' (for example *Cibicides lobatulus*) to average life-spans of 3 to 5 months. Similar life-spans are supported by the Kosterfjord results with a first colonisation of *Cibicides lobatulus* within 6 months from late autumn to early spring and the considerably increased recruitment after 12 months comprising additionally spring and summer. Therefore, the carbonate accretion rates of *Cibicides lobatulus* in the Kosterfjord experiment represent a minimum estimate since reproduction rates or the number of abandoned attachment scars on the carbonate plates is not taken into account.

The accretion rate by *Cibicides lobatulus* in the second year of the experiment stays equal at the 7 m station but is slightly lower at the other stations. Only at 30 m a drastic decrease occurred with values 15 times lower compared to the first year (Table 12). This finding may be the result of more unfavourable conditions in terms of the hydrologic setting or nutrition supply during the period from October 2003 to September 2004 as compared to the period from October 2002 to October 2003.

Limits for carbonate production are considered by Wefer & Lutze (1978) to be mechanical stress with strong turbulence at shallower depths preventing the growth of larger species with higher standing stock as well as low food availability and the extreme seasonality of water mass properties. In the case of the present study, these limitations are of minor importance. The close correlation between maximum carbonate accretion of *Cibicides lobatulus* and the mixed water mass layer may indicate that *Cibicides lobatulus* is perfectly adapted to varying environmental conditions including temperature, salinity, current strength and turbidity.

The difference of carbonate accretion between the two substrate types is seemingly a result of substrate preference of the individual species (see above). However, even higher accretion rates would have been observed on the limestone plates when taking the number of attachment scars of *Cibicides lobatulus* into account (for example 80,000 scars per square metre at 15 m after two years). These prominent attachment scars witness the former presence of benthic foraminiferans which have long been vanished. On fossil hardgrounds, they may thus serve as an in situ proxy for the presence and abundance of foraminiferans. The epibenthic foraminiferal tests themselves are rarely if at all preserved in life position and are predominantly transported and incorporated as allochthonous components in nearby soft bottom sediments. In general, transport of dead benthic foraminiferal tests by currents is regarded as substantial (Alve 1999). This process becomes already apparent during the course of the present 2-year experiment, where the attachment scars recorded on the carbonate plates clearly outnumber the total foraminiferal specimen encountered – often by a more then tenfold abundance. As an additional explanation for the higher number of attachment scars if compared to foraminiferal tests, migration behaviour leading to several attachment scars produced during the life-span of one individual is to be considered. A related behavioural pattern, namely the development of organic cysts during resting phases of some bathyal benthic foraminiferans has been observed during laboratory experiments conducted by Gross (2000) and specifically *Cibicides lobatulus* was characterised as temporarily motile by Langer (1993).

Substrate orientation

The orientation of the substrate is known to influence the settlement intensity of encrusting organisms with accretion rates usually being significantly higher on the exposed downward facing compared to the upward facing side of the substrate. Reaka-Kudla et al. (1996) for instance recorded a significantly higher carbonate accretion on the bottom side of planted micrite blocks than on their top side (70 versus 350 g/m^2/y at 5-6 m and 121 versus 614 g/m^2/y at 11 to 13 m water depth, respectively) during their experiments at the Galápagos Islands as did Maughan (2001) during his experiments on the influence of sedimentation and light on epifaunal recruitment in SW Ireland. A number of reasons for this orientation dependence can be addressed as there are: (1) preference of shaded bottom side, (2) enhanced grazing pressure and mortality of encrusters on the top side, and (3) enhanced background sedimentation on the top side.

In the present study, this observation could be qualitatively confirmed by visually inspecting the bottom and top sides of the experimental panels. Especially serpulimorphs and bryozoans were much (roughly 5 to 10 times) more abundant on the bottom side. This observation holds true for all depths from 7 m to 85 m. The 1 m panels were directly mounted on a large concrete block prohibiting a corresponding comparison. The assessed carbonate accretion rates are referring exclusively to upward facing substrates and can thus be considered as minimum estimations in the given environment.

Carbonate accretion rates and bathymetry

The bathymetric dependency of carbonate accretion rates is not as clearly expressed as in the the case of the bioerosion rates. There is, nevertheless, a pattern of an increase towards 15 m water depth, owing to the highest population densities of many epizoan groups such as foraminiferans, Serpulidae and Escharellidae. At the 30 and 50 m stations, very low accretion rates are exhibited and at 85 m, the rates are somewhat enhanced due to the proximity to the Säcken Reef site. This pattern is overprinted by the mass occurrence of balanids at the shallowest station, resulting in the by far highest rates encountered during the experiment, exceeding all other values by several hundred times. The balanids, however, were only encountered at the transect and not at the corresponding lagoonal site. In summary, we deal with a heterogeneous bathymetric dependency with a distinct maximum at 1 m water depth and a subordinate maximum at 15 m.

Carbonate accretion and exposure time

There is no constant pattern concerning the link of carbonate accretion rates with exposure time. There is a significant increase from zero to the highest rates recorded during the experiment at the 1 m station between 6 and 12 months exposure owing to mass occurrence of balanids. The 24 months panels from this depth were lost, but it can be expected that the rates would have decreased somewhat since almost all free settlement space was occupied by these epizoans already after 12 months exposure. When excluding this depth from the evaluation, a constant to weakly increasing tendency towards longer exposure times can be seen at most depths. This pattern is possibly grounded in a weak competition for space on the generally sparsely colonised substrates in combination with the facilitation of larval recruitment after biofilms have developed.

7 Ecological and palaeoenvironmental implications

7.1 Bathymetry

In Sect. 4.6 we have seen, that when applying the established palaeobathymetric index ichnocoenoses scheme (see Sect. 1.7) to the present material, the general trends in bathymetric distribution of the various trace maker groups and their corresponding traces can be clearly confirmed, while some of the individual index ichnocoenoses were of limited applicability. As a consequence, the very common trace *Cavernula pediculata* with its cosmopolitan biogeographic distribution was suggested as a substitute for the even in the Tropics rare ichnospecies *Fascichnus acinosus*, and '*Fascichnus*-form 1' for '*Palaeoconchocelis starmachii*', respectively. The ichnotaxon *Cavernula pediculata* is known at least back to Triassic times (Glaub & Vogel 2004) while the stratigraphical range of '*Fascichnus*-form 1' is not known yet. The dysphotic ichnocoenosis is clearly indicated by *Ichnoreticulina elegans* being the only trace produced by a phototrophic organism. With this flexibility at hand, the value of microendolithic ichnocoenoses for judging relative (palaeo)bathymetry is confirmed. The ichnospecies *Ichnoreticulina elegans* (trace of the green alga *Ostreobium quekettii*) serves as an appropriate indicator for the photic limit as it does in the Tropics. Rhodophyte *Conchocelis*-stages were of no corresponding value but may well be encountered in other cold-temperate to polar settings.

With the more and more condensed photic zonation towards higher latitudes – as a result of weaker sunlight radiation and considerable seasonal eutrophication – the shallow-subtidal photic subzones may not be as clearly reflected in the ichnocoenoses (especially in areas with macrotidal regimes). The specific ichnocoenoses of the inter- and supratidal zones on the other hand, should be promoted by enhanced tidal movements.

7.2 Latitude and temperature

The biogeographic distribution pattern of endoliths reflects a general trend of depletion in the phototrophic borer spectrum with higher latitude as a result of the seasonal fluctuation of light availability (Schmidt & Freiwald 1993) and significant differences concerning their lower temperature resistance limit (Lüning 1985). As a consequence, forms missing in cold-temperate and/or polar waters bear a value as palaeotemperature and -latitude indicators. In this respect, the record of a number of taxa in the Kosterfjord experiment is extending their known biogeographic distribution northward. The trace *Eurygonum nodosum* for instance was absent in other North Atlantic settings (Spitsbergen, Tromsø, W Scotland) and Glaub et al. (2002) consequently concluded that its trace maker – the cyanobacterium *Mastigocoleus testarum* – favours warmer waters. However, the original description by Lagerheim (1886) reports *Mastigocoleus testarum* from Swedish waters and Nielsen (1972) found it at the island of Læsø (Denmark), the two northernmost occurrences reported to date. The occurrence of *Mastigocoleus testarum* in the Kosterfjord shifts the northern limit of its biogeographic distribution slightly farther north. The same reasoning applies for *Fascichnus* cf. *acinosus* (trace maker *Hyella balani*). The trace *Fascichnus grandis* (produced by the dasycladacean *Acetabularia*) does not occur in the present material and was also found to be absent in other North Atlantic sites (Glaub et al. 2002). Utilizing the biogeographic distribution with a limit in the cold-temperate realm for the latter species, the occurrence of *Eurygonum nodosum*, *Fascichnus grandis* or *Fascichnus acinosus* in fossil carbonates is circumstantial evidence against a polar setting.

On the other hand, the ichnospecies *Rhopalia catenata* was reported as far north as the Tromsø area, Norway (Glaub et al. 2002) and their two potential trace makers, the green algae *Eugomontia sacculata* and/or *Phaeophila dendroides*, were reported from northern Scotland (Akpan & Farrow 1984a), Norway (Rueness et al. 2001) and from the Faroes (Nielsen & Gunnarsson 2001). The northernmost occurrence of *Eugomontia sacculata* was reported from Iceland (Caram & Jónsson 1972), reflecting its ability to cope with lower temperatures than the species discussed previously. The same applies to *Epicladia testarum*, the assumed trace maker of *Eurygonum pennaforme* which is known from Danish and Swedish waters (Nielsen 1972; Nielsen et al. 1995) and from Norway (Rueness et al. 2001). The highest temperature tolerance of all chlorophytes have *Gomontia polyrhiza* (trace: *Cavernula pediculata*) and *Ostreobium quekettii* (trace: *Ichnoreticulina elegans*) which both thrive well in the Tropics to the Arctic (Taylor 1957; Lund 1959; Pedersen 1976). They are consequently of no use as palaeotemperature indicators but well suitable as global bathymetric key ichnotaxa. The same applies for *Fascichnus dactylus*, which can be produced homeomorph by a number of different cyanobacteria some of which being cosmopolitan, as well as for *Scolecia filosa* and *Planobola*.

Endolithic rhodophytes or their traces were neither found in the experimental substrates nor in the background sample material. However, this is a group which has been largely ignored by ichnotaxonomists so far and consequently not a single ichnotaxon has been established ('*Palaeoconchocelis starmachii*' is not a clearly defined ichnotaxon even though it has often been applied as such).

Endolithic fungi bear a considerable potential as palaeotemperature indicators but are poorly studied in this respect yet. While the traces *Orthogonum lineare*, *Saccomorpha clava* and '*Flagrichnus*-form 1' have a cosmopolitan distribution and are consequently of no indicative value, *Saccomorpha terminalis* and its producer, the fungus *Phytophthora* for instance have preferentially been encountered in cold-temperate to polar settings and further assessment of its biogeographic distribution could yield a corresponding suitability as palaeotemperature indicator. The same applies to '*Flagrichnus*-form 2', which to date has never been reported from Recent or fossil tropical or subtropical settings, despite a wealth of literature dealing with microendoliths from those settings. Indeed, this trace was exclusively encountered in Recent cold-temperate to cool-water settings with water temperatures down to just above freezing (see Wisshak & Porter in press, and a table of occurrences therein). The only related Recent low-latitude occurrence (Florida Escarpment) stems from cold bottom waters (~4.5°C; Paull et al. 1984) in more than 3,000 m water depth (Hook & Golubic 1993) reminding us, that we must not confuse low (palaeo)temperature with high latitude, in spite of a certain degree of correlation. The two Pleistocene holotype occurrences reported by Wisshak & Porter (in press) and the only record in open nomenclature recently given in the literature by Vogel & Marincovich (2004) from the Tertiary of Alaska stem from non-tropical settings. This finding leads to the assumption, that its unknown fungal producer has a low upper temperature tolerance limit, making '*Flagrichnus*-form 2' another prime candidate as indicator for low (palaeo)temperatures. Both *Saccomorpha terminalis* and '*Flagrichnus*-form 2' potentially indicate cool- to cold-water conditions by their presence and are consequently of especial value as opposed to palaeotemperature interpretations based on the absence of certain taxa (see above) which are based on negative evidence and are thus of limited reliability.

7.3 Salinity and temperature

Another conclusion that can be drawn from this experiment concerns the salinity and temperature tolerance of those species thriving well at the shallowest stations. There, they have to cope with considerable fresh-water input leading to salinity values often lower than 20 psu (occasionally dropping to less than 10 psu) and temperature fluctuations ranging from 0°C (sea ice) to 20°C. At the 7 m station, salinities are also quite low, frequently dropping below 25 psu, again coupled with strong temperature fluctuations. Consequently, most phototrophs encountered can be regarded as considerably euryhaline and eurytherm, respectively. However, some traces might have been produced during a relatively short time period under a narrow range of hydrographic changes as for instance a stable warm spell during summer time. This factor can only be excluded by cultivation experiments, implying the need for future studies in this field. Nevertheless, many cyanobacteria species, such as *Hyella caespitosa*, are known even from pure freshwater environments (Schneider 1999). Reasoning the other way around, most fungal endoliths (except for *Ostracoblabe implexa*; see below) appear preferentially from 15 m on downwards indicating their need for more stable marine hydrographic conditions or less competition by phototrophs.

The fungus *Ostracoblabe implexa* (trace: *Orthogonum fusiferum*), which was almost exclusively encountered at the lagoonal setting and in the *Littorina* shells, is known as obligate marine with optimum growth at 23 to 25 psu, but tolerating lower salinities under considerably reduced growth rates (Alderman & Gareth Jones 1971). In temperatures below 15°C, growth is almost inhibited giving an explanation for its preference for the shallow-water and especially the lagoonal setting, where summer temperatures exceed those at the transect. The same authors showed, that growth was significantly reduced under illuminated conditions, explaining their absence in the Iceland spar and the deeply penetrating habit of the trace in the *Callista* and *Littorina* shells. Consequently, a deeply boring habit and the appearance of this trace together with cyanobacterial and chlorophyte traces may provide circumstantial evidence for a fossil shallow and temperate setting.

7.4 Interpretation of cold-water coral occurrences

The colonial, habitat forming azooxanthellate scleractinian coral *Lophelia pertusa* is regarded as a typical member of the bathyal zone community encountered especially in the Atlantic Ocean and Mediterranean Sea along deep shelves, oceanic banks, seamounts and continental margins. This zone as a marine ecologic realm extends from the edge of the continental shelf to the depth at which the water temperature drops below 4°C. Both of these limits are variable, but the bathyal zone is generally described as lying between 200 and 2,000 m below sea level, or as the aphotic zone where only heterotrophic and chemotrophic organisms can exist. This broad environmental definition matches well with the majority of the known occurrences of *Lophelia pertusa* (see Rogers 1999). Based on this knowledge, ancient *Lophelia*-bearing deposits are regarded by palaeontologists as deep-water or as bathyal (Squires 1964; Barrier et al. 1989; Hanken et al. 1996; Tischack & Freiwald 2005).

In the Swedish Kosterfjord area, however, we deal with non-bathyal *Lophelia pertusa* occurrences, well exemplified by the Säcken Reef site, where corals thrive at only 80-90 m water depth. Such a finding is not unique for *Lophelia pertusa*, as even shallower locations are known from Norwegian inshore and fjord environments (Dons 1944; Strømgren 1971; Fosså et al. 2000, 2002). Are these shallow-water *Lophelia* occurrences "extreme", or have scientists underestimated the ecological capacity of this coral species? The *Lophelia* occurrence of the Säcken Reef site persists only a few tens of metres beneath a permanently brackish surface water layer. However, at the depth of the coral patches, the hydrographic data reveal fully marine conditions (Sect. 3.2), which are ensured by a deeper inflow of Atlantic water through the more than 700 m deep Norwegian Trench into the Skagerrak to compensate the surface water net outflow. Similar situations are met in some Norwegian fjords, where live *Lophelia* reefs are known to exist at much shallower depths compared to the open shelf off Norway (Freiwald et al. 1997). To conclude, non-bathyal, shallow-water *Lophelia* occurrences in Scandinavian waters are confined to areas, where saline oceanic waters can intrude as topographically-guided underflows onto the inner shelf and adjacent fjords, driven by an estuarine circulation combined with tidal currents. Hence, environmental conditions in the Kosterfjord are not "extreme" and existing knowledge about the ecological limits of *Lophelia pertusa* is still valid.

The endolithic assemblage and respective ichnocoenosis encountered in *Lophelia* skeletons from the Säcken Reef site (Sect. 4.3), is dominated by traces produced by boring sponges (trace: *Entobia*), the bryozoan *Spathipora* (trace: *Pennatichnus*), the fungus *Dodgella priscus* (trace: *Saccomorpha clava*) and an unknown fungus (trace: *Orthogonum lineare*). The relative abundance of all further traces, most of which are here tentatively assigned to the boring activity of micro-sponges or unknown fungi, is low. This ichnocoenoses, composed solely of traces produced by heterotrophs, resembles the *Saccomorpha clava* / *Orthogonum lineare* ichnocoenosis, which is regarded as indicative for fossil and Recent open marine, aphotic environments (see above). Hence, from an ichnological point of view, the present endolithic assemblage corresponds well to ichnocoenoses from other, but real bathyal North Atlantic settings, as reported for instance by Schmidt & Freiwald (1993) from shell material off northern Norway (170 m water depth), Boerboom et al. (1998) in skeletons of the deep-sea coral *Desmophyllum christagalli* from Orphan Knoll (>1,700 m water depth) and Beuck & Freiwald (2005) in *Lophelia* skeletons from the Propeller Mound (Porcupine Seabight, >680 m water depth).

This finding bears some important palaeoecological implications. Provided that the non-bathyal coral community will become fossilised, geologists would encounter difficulties finding convincing arguments for a non-bathyal environment. These difficulties arise from the fact, that the conclusions were drawn from real bathyal *Lophelia* communities of the open marine deep-water realm. The finding of bathyal communities in comparatively shallow waters is linked to factors that force deeper oceanic water masses to surface. Such situations are likely to be expected where an estuarine circulation prevails. Other scenarios are deep-sea basins bordered by narrow shelves and with local upwelling cells driven by the wind regime, facilitating the intrusion of eutrophic deeper waters to shallow depths – together with the benthic communities. Areas of interest, where such a combination of factors may occur, are island arcs with a steep bathymetric gradient towards the subduction zone and narrow straits. Strikingly, almost all known exposed ancient *Lophelia* locations derive from tectonically active regions with steep bathymetric gradients and a specific confined topography, which could have forced deep-water to the near surface. Corresponding candidates are Rhodes as part of the Hellenic Arc in the eastern Mediterranean Sea (Hanken et al. 1996; Bromley 2005; Titschack & Freiwald 2005), occurrences in the Messina Strait (Barrier et al. 1989) and the Cook Strait, New Zealand (Squires 1964).

7.4 High-latitude versus low-latitude bioerosion

Concerning the pace and rate of bioerosion in the cold-temperate Kosterfjord compared to tropical reef settings, we have seen in Sect. 6.2, that tropical rates exceed the high-latitude ones by 2 to 10 times when considering exclusively microbioerosion and by 10 to 100 times in respect to overall bioerosion rates. This finding strongly suggests a general latitudinal gradient. The decline is on the one hand linked to the lower ambient temperatures in concert with a restricted light availability, slowing down (bio)chemical processes and activity. On the other hand, specific characteristics in the composition of the bioeroding fauna and flora promote these quantitative differences.

A moderate depletion in microborer diversity, especially in terms of chlorophytes (see above) can be observed between the Tropics and the present study site – and can be expected to be even more pronounced towards polar waters. While few traces such as *Fascichnus grandis* are restricted to low latitudes, *Eurygonum nodosum, Fascichnus acinosus*, *Rhopalia catenata* and *Eurygonum pennaforme* are found up to cold-temperate waters and cosmopolitan ichnotaxa are represented for instance by *Cavernula pediculata, Ichnoreticulina elegans, Fascichnus dactylus, Scolecia filosa, Orthogonum lineare, Saccomorpha clava* and '*Flagrichnus*-form 1'. Only few ichnotaxa such as *Saccomorpha terminalis* and '*Flagrichnus*-form 2' are in turn restricted to cold-temperate or polar high latitudes.

In terms of macroborers, all principal groups of bioerosion agents found in the Tropics are also present in high latitudes, albeit mostly replaced by different taxa. Some of the most effective borers in tropical to warm-temperate settings, however, lack comparably effective relatives in higher latitudes as for instance the prominent boring mussel *Lithophaga*. Other groups which are especially relevant and ubiquitous in the Tropics are negligible in high-latitude settings as for instance the sipunculans. Yet other groups are in turn more relevant in cold-temperate to polar waters such as the various macroalgae.

Major differences concern the relative importance of grazers in both latitudes. Especially significant is the almost complete absence of grazing herbivorous fish (scarids, acanthurids, tetradontids) in favour of far less effective herbivorous invertebrates (Menge & Lubchenco 1981; Horn 1989). These parrotfish and triggerfish account for the highest relative percentage of bioerosion in tropical reef settings in concert with the bioerosion force of grazing echinoids. Grazing rates determined by the experiments in the various tropical reef settings often exceed several -1,000 g/m^2/y (see Table 1) and the highest rates recorded amount to an impressive -10,380 g/m^2/y (Peyrot-Clausade et al. 1995b) recorded in French Polynesia and -22,800 g/m^2/y determined by Reaka-Kudla et al. (1996) at the Galápagos archipelago. The only relevant fish grazers in the Kosterfjord area are the labrids, most numerous of which are *Ctenolabrus rupestris*, *Symphodus melops* and subordinately *Labrus berggylta* and *Labrus bimaculatus* (Lundälv & Jonsson 2000; Lundälv pers. comm.). No traces attributable to these fish were encountered during this study and it is possible that – unlike the tropical parrotfish – these species graze without considerably affecting the hard substrate. Concerning the echinoids, the most important species in shallow waters is *Psammechinus miliaris*, while in somewhat deeper waters, occasionally high numbers of *Echinus esculentus* and *Strongylocentrotus droebachiensis* may be encountered (Lundälv & Jonsson 2000; Lundälv pers. comm.). Even though all of these species are bioerosion agents in high latitudes (see Table 2), their traces were encountered only sporadically during the present study. Their overall erosive force certainly lacks behind that of their tropical counterparts such as *Diadema antillarum*, mass occurences of which govern the bioerosion rates in some sites, as reflected for instance in the -5,300 g/m^2/y reported by Scoffin et al. (1980b) from a Barbados reef.

For the degree of inter- to supratidal bioerosion, however, a contrary latitudinal trend might be drawn. Kelletat (1988), comparing results of his various studies on this topic in mid- and high-latitudes in both hemispheres (including quantitative investigations), found a raised littoral bioerosion activity in cool and cold climates. Schneider & Torunski (1983) on the other hand – summarising results of previous quantitative studies in this respect – found that erosion rates were higher in the summer months than in the winter. They considered the reason in the variations of ambient temperature, which implies lower rates towards colder climates in higher latitudes.

7.5 The preservation potential of high-latitude carbonates

On a first sight, lower bioerosion rates towards lower temperatures and higher latitudes, respectively, imply a superior chance of cold-water skeletal carbonates to become preserved in the fossil record. This stands in opposition to the observation that calcareous skeletons retrieved from cold-temperate to polar seafloors often show especially strong signs of bioerosion (e.g., Farrow & Fyfe 1988). This is not only the case in the shallow-water euphotic zone, where bioerosion is most promoted, but also in aphotic depths. Cold-water coral reefs for instance – the key carbonate factories in high-latitude seas – are often intensely degraded by bioeroders (see Sect. 4.3; Freiwald & Wilson 1998; Beuck & Freiwald 2005). The solution for this apparent contradiction is found in lower sedimentation rates combined with much longer retention periods of skeletal carbonates exposed on the seafloor for hundreds to thousands of years (Smith & Nelson 2003). Nevertheless, a fair amount of skeletal debris becomes incorporated in the accumulating sediment, leading to a net build-up of for instance deep-water coral reefs and carbonate mounds (e.g., Lindberg & Mienert 2005). The major portion of the skeletal carbonate is, however, either dissolved or broken down to silt and mud-sized grains by the action of the various bioerosion agents in concert with mechanical abrasion, and subsequently become a component of carbonate sediments on the shelf (Farrow & Fyfe 1988).

What can we learn in this context from the Kosterfjord experiment? For instance, we have noticed that at all depth stations, physicochemical dissolution is negligible for shells exposed on the seafloor. This holds true not only for the experimental substrates deployed for two years, but also for the investigated taphonomically-advanced background sample material (Sects. 4.3-4.4). Clearly, bioerosion is the chief process driving the degradation of skeletal carbonates in this setting, confirming the observations of Farrow & Fyfe (1988), Cutler & Flessa (1995) and many others, and opposing the view of Alexandersson (1979), who considers the physicochemical dissolution as the main destructive process. Moreover, carbonate precipitation above the sediment/water interface, other than the weak accretion by calcareous epizoans, is not a relevant factor in the present setting. This stands in contrast to the widespread early diagenetic development of calcite biospar formed within abandoned microborings repeatedly observed in tropical and subtropical seas (Bathurst 1966; Smith 1988). Inorganic precipitation and marine cementation are generally absent or slow in cool waters (Smith & Nelson 2003), even though exceptions occur (Farrow & Fyfe 1988). As a consequence, an enhanced mechanical stability owing to constructive diagenesis in the tropical and subtropical seas is lacking, further promoting the breakdown. Cutler & Flessa (1995) nevertheless recognised a certain degree of "case-hardening" in the intertidal shells they studied from the Skagerrak.

As the bottom line, effects of early seafloor processes on the taphonomy of skeletal carbonate deposits in cold-temperate to polar seas form a complex interplay of various chemical (dissolution, precipitation), physical (abrasion, breakage) and biological (bioerosion, encrustation, bioturbation) processes, differentially determining the fate of cold-water carbonates (Smith & Nelson 2003). Owing to the long residence of skeletal carbonates exposed on the seafloor as a result of low sedimentation rates, bioerosion – even though at a significantly slower pace than in the Tropics and subtropics – is a strong force in high-latitude seas, underlining the comparatively low overall preservation potential of cold-water carbonates.

8 Summary and conclusions

In the following account, the main findings of the Kosterfjord experiment are summarised, following the consecutive order of Sects. 3 to 7, before topping off the present volume with a brief outlook in Sect. 9.

8.1 The Kosterfjord study site

- The Kosterfjord lies in the northeastern part of the Skagerrak, which is the major gateway between the North Atlantic and the Baltic Sea, where the overall oceanographic regime is driven by an estuarine circulation pattern. Incoming dense, saline and oxygenated oceanic water underflows the more brackish surface water outflow of the Baltic Sea. This watermass stratification is clearly expressed most of the year with a thermocline and halocline at around 20 to 40 m water depth. A thin low-salinity top layer is added by local river runoff. This stratification is most prominent in the coldest (February and March) and warmest (July and August) months in respect to the sea surface temperature (SST).
- The SST is subject to strong seasonal fluctuations, ranging from below freezing (sea ice) in the coldest, to 17-20°C in the warmest months. This annual fluctuation is also present in deeper waters, but with lower amplitude and a certain time delay of several months. Freshwater input leads to strong salinity variations of the surface waters from 8.0 psu to 30.9 psu with the lowest values present between May and August. The oxygen level ranges from 4.2 to 9.5 ml/l with the highest values found in springtime.
- The Säcken Reef site, located on a sill in the northern Kosterfjord area, comprises two major patches with living *Lophelia pertusa* cold-water corals (250 m^2 and 50 m^2 in size), situated on the flanks of two small mounds in around 85 m water depth. Most of the colonies are small and grow as discrete, often spherical colonies with lumps of dead coral, coral rubble, gravel, cobble and sediment in between. There, *Lophelia pertusa* thrives under fully marine conditions, with temperature fluctuations between 4 and 10°C (mean 6°C), salinity fluctuation between 33.6 and 35.0 psu (mean 34.5 psu) and oxygen fluctuation between 4.8 and 6.7 ml/l (mean 6.1 ml/l). This physical oceanographic spectrum matches well the conditions found at *Lophelia* sites along the northwestern European continental margin.

- Short-term fluctuations of watermass properties were investigated by two lander deployments near the Säcken Reef site. The temperature at this depth is subject to only minor fluctuations of less than one degree and salinity variations are in the range of 0.4 psu. The turbidity was found relatively constant at around 0.5 ntu and the current velocity occasionally reached close to 25 cm/s, with a mean of 5.3 cm/s.

- Owing to the high geographic latitude (59°) combined with dominating eutrophic conditions, a condensed photic zonation is encountered. Direct measurements of the photosynthetically active radiation (PAR) under various weather conditions and different seasons indicate a base of the euphotic zone (drawn at 1% surface illumination) in 10-25 m water depth and a base of the dysphotic zone (~0.01% surface illumination) ranging in 30-60 m water depth. At the Säcken Reef site in 85 m water depth, aphotic conditions prevail.

8.2 Bioerosion patterns

- The microendolithic inventory (trace diameter <100 μm) of the experimental substrates, studied by SEM analysis of epoxy resin casts, yields diverse ichnocoenoses comprising a total of 21 traces produced by boring cyanobacteria (*Eurygonum nodosum*, *Planobola*, *Scolecia filosa*, *Fascichnus* cf. *acinosus*, *Fascichnus dactylus*, *Fascichnus frutex* and '*Fascichnus*-form 1'), chlorophytes (*Cavernula pediculata*, *Ichnoreticulina elegans*, *Rhopalia catenata* and *Eurygonum pennaforme*), fungi (*Saccomorpha clava*, *Planobola radicatus*, *Saccomorpha terminalis*, '*Flagrichnus*-form 1', '*Flagrichnus*-form 2' and *Orthogonum fusiferum*) and traces of uncertain affinity (*Orthogonum lineare*, '*Orthogonum*-forms 1-2' and *Scolecia serrata*).

- The inventory is complemented by traces studied in background samples from the two bathymetric end members of the transect, namely skeletons of the cold-water coral *Lophelia pertusa* sampled at the Säcken Reef site and the shallow-water gastropod *Littorina littorea* collected at the shoreline. While the experimental substrates document the early to intermediate stages of endolithic colonisation, the background samples record advanced taphonomical stages. Only the bryozoan traces *Pennatichnus* and *Iramena* as well as '*Orthogonum*-form 3' and 'Problematic-form 1' were exclusively encountered in the aphotic background samples and *Fascichnus rogus* in the shallow-euphotic material, increasing the total number of recorded microboring ichnotaxa to 26.

- The relationship between the endoliths (biotaxa) and the traces they leave (ichnotaxa) is evaluated by the study of the boring organisms in situ by transmission light microscopy of planted Iceland spar and translucent bivalve shells. In total, 18 endolithic taxa are recorded, comprising cyanobacteria (*Mastigocoleus testarum*, *Plectonema terebrans*, *Hyella balani*, *Hyella caespitosa*, *Solentia foveolarum*, *Solentia achromatica*, *Hyella gigas*, *Kyrthutrix dalmatica*, *Scytonema endolithicum*), chlorophytes (*Gomontia polyrhiza*, *Ostreobium quekettii*, *Phaeophila dendroides*, *Eugomontia sacculata*, ?*Epicladia testarum*), and fungi (*Dodgella priscus*, *Phytophthora*, *Ostracoblabe implexa*).
- Traces of macroborers (trace diameter >100 µm) and grazers are of minor importance during early stages of colonisation but were noteworthy diverse and abundant in the experimental substrates after 2 years exposure and in the background sample material, complementing the inventory to a total of 18 ichnotaxa. The recorded traces comprise various sponge borings (*Entobia*, 'Microsponge-forms 1-6'), polychaete borings (*Caulostrepsis taeniola* and *Caulostrepsis cretacea*), predation traces of echinoids (*Gnathichnus pentax*), grazing traces of chitons and/or gastropods (*Radulichnus inopinatus*), attachments scars of anomiid bivalves (*Centrichnus eccentricus*), possible borings of foraminiferans ('*Semidendrina*-form'), brachiopod attachment scars (*Podichnus centrifugalis*) and one unidentified trace ('Problematic-form 2'). Shallow attachment scars chiefly produced by *Cibicides lobatulus* ('Foraminiferan-form 1') represent by far the most abundant macroboring with a maximum density of more than 80,000 scars per square metre (15 m after two years exposure). A previously unknown specific attachment scar ('Foraminiferan-form 2') was found to be produced by the rare *Gypsina vesicularis*, and the parasitic *Hyrrokkin sarcophaga* signs responsible for 'Foraminiferan-form 3'.
- The experimental approach allows the observation of relative order and pace of colonisation by microborers in relation to light availability and bathymetry: at the 1 m station, a mature shallow-euphotic ichnocoenosis (dominated by cyanobacteria) and at 7 m, a deep-euphotic ichnocoenosis (dominated by chlorophytes), respectively, is developed already after 12 months exposure. With the vanishing light from 15 m downwards, the ichnocoenoses development is significantly slowed down and only immature dysphotic and aphotic borer communities (dominated by fungi) are encountered.

- The highest diversity with respect to microbioerosion is found at 7 m water depth (12 ichnotaxa), followed by the 1 m station (10 ichnotaxa). With only 5 ichnotaxa, the lagoonal setting is considerably depleted in ichnotaxa richness. As a result of the vanishing light towards greater depths, the diversity decreases to only 5 ichnotaxa at the 50 m station and slightly increases in the vicinity of the cold-water coral reef (7 ichnotaxa). A similar pattern is developed for the corresponding endoliths recorded in Iceland spars and mollusc shells, as well as for ichnotaxa recording the activity of macroborers.
- Distinct differences, specifically in the abundance and diversity of macroboring traces, are expressed when comparing the very initial ichnocoenoses encountered in the experimental substrates at the aphotic Säcken Reef station with the mature ichnocoenoses recorded in the background sample material from the same site. Sponge borings were limited to few initial *Entobia* cavities in the experimental substrates but represent the dominant traces in the *Lophelia* skeletons. Traces produced by boring bryozoans (*Pennatichnus*, *Iramena*) were exclusively and commonly encountered in the background samples. The two opportunistic '*Flagrichnus*-forms' and *Saccomorpha terminalis* were in turn only encountered in the experimental substrates.
- The results obtained from the shallow-euphotic experimental substrates well mirror the ones recorded from the background *Littorina littorea* samples even though the latter being a mobile substrate of different shape and ultrastructure. Except for *Fascichnus rogus*, all ichnotaxa recorded in the *Littorina* shells were also found in the experimental substrates with the same dominating taxa.
- The illumination status, recorded by measurements of the photosynthetically active radiation (PAR), is confirmed by the bathymetric distribution range of phototrophic endoliths, with a deepest record of the chlorophyte *Ostreobium quekettii* (trace: *Ichnoreticulina elegans*) at the 30 m station and a limit of 15 m for all other phototrophs.

8.3 Carbonate accretion patterns

- Calcareous epizoans encountered on the experimental substrates comprise bryozoans (Tubuliporidae and Escharellidae), serpulimorphs (Serpulidae and Spirorbidae), balanids (Balanidae and Verrucidae), crinoids (Antedonidae) and a diverse fauna of epibenthic foraminiferans.

- The foraminiferal assemblage comprises 12 species, 8 of which are belonging to the Rotaliina and 4 to the Textulariina. The most abundant species are *Cibicides lobatulus* (51.24%), *Lituotuba lituiformis* (37.51%), *Nubecularia lucifuga* (6.52%), *Planorbulina mediterranensis* (3.24%) and *Tholosina vesicularis* (1.11%). All others show an accessory contribution to the assemblage with less than 1%. All encountered species except for *Nubecularia lucifuga* and *Gypsina vesicularis* have previously been reported from Scandinavian waters, but within different assemblage compositions.

- The calcareous epibenthos is not evenly distributed with water depth and exposure time, but shows characteristic distribution patterns. The overall bathymetric distribution means clearly express the dominance of small epibenthic foraminiferans with population densities reaching many thousands of *Cibicides lobatulus* and *Lituotuba lituiformis* per square metre. Solely the balanids reach comparable numbers at the shallowest station, where they outcompete all other epizoans. Serpulimorphs appear in considerable numbers, while the bryozoans are comparatively rare and crinoid rhizoidal parts were only found at the deepest station. Whereas most epizoans show their distribution maximum in the shallower waters in 7 to 15 m water depth, the Spirorbidae, Tubuliporidae and Antedonidae increase in abundance towards deeper waters.

- A heterogeneous pattern is drawn by the bathymetric distribution for different exposure times on the PVC settlement plates. Only few families such as the Tubuliporidae and Escharellidae show a steady increase in specimen numbers with time and distinct bathymetric maxima, while others decrease with time (Antedonidae and Spirorbidae) or show a completely heterogeneous pattern (Serpulidae). The five dominant foraminiferal species, however, show a strong increase with time and a pronounced maximum at the 15 m station with a population density of close to 80,000 individuals per square metre after two years exposure. At all depth stations, the foraminiferal assemblage is dominated by *Cibicides lobatulus* and *Lituotuba lituiformis*. The maximum species diversity is encountered at intermediate water depths in 15 to 30 m with 10 and 9 species, respectively. In the shallower waters, specimen and species numbers are depressed by the strong hydrologic variability and enhanced spatial competition with other biota.

- When comparing the population densities of the various epizoans on the limestone versus the PVC plates, specific substrate preferences become apparent: Serpulidae and Balanidae are moderately more abundant on the carbonate substrates, the Spirorbidae and Tubuliporidae show no significant preference, and the Escharellidae and Antedonidae favour the PVC substrates. The various foraminiferans have differential preferences with *Cibicides lobatulus, Planorbulina mediterranensis* and *Tholosina vesicularis* clearly favouring PVC whereas *Lituotuba lituiformis* is about equally distributed on both substrate types and *Nubecularia lucifuga* prefers the calcareous substrates.

8.4 Quantitative bioerosion and carbonate accretion

- The gravimetrically determined bioerosion rates range from a mean of close to zero at the deepest station to a maximum of -218 g/m^2/y (lagoon after 6 months exposure) with a mean of -37 g/m^2/y. The standard deviations are moderate to high. Highest bioerosion rates are found at the shallowest stations and especially in the lagoonal setting, while with one exception the values at all deeper stations did not exceed -50 g/m^2/y. At most depths, bioerosion rates were decreasing with time which is also expressed in a decrease in overall means per exposure time at the transect from -54 g/m^2/y (6 months) to -43 g/m^2/y (12 months) to -35 g/m^2/y (24 months; assuming constant rates at the lost 1 m station).
- The respective carbonate accretion rates range about two magnitudes lower from zero to 1.29 g/m^2/y, with one extreme exception, represented by the 1 m station after 12 months, where massive encrustation by balanids accounts for 361.93 g/m^2/y, resulting in an overall mean of 19 g/m^2/y. The standard deviations are high and often exceed the mean values. There is an overall increase in mean accretion values with time (untypical shallowest station excluded) from 0.26 g/m^2/y (6 months) to 0.36 g/m^2/y (12 months) to 0.49 g/m^2/y (24 months). The limestone plates exhibit slightly higher overall accretion rates (20.99 g/m^2/y) if compared to the PVC plates (17.89 g/m^2/y).
- Estimated carbonate accretion rates for *Cibicides lobatulus* (based on the mean test weight and population densities) range from 0 to 0.326 g/m^2/y with the highest rates occurring at 7 to 30 m water depth after the first year of exposure. Accretion rates are up to two magnitudes higher on the PVC than on the carbonate substrates. The rates represent a minimum estimate by not taking reproduction rates and the number of abandoned attachment scars into account but are nevertheless up to two magnitudes higher than estimates previously reported from the western Baltic Sea.

- The significant decrease in bioerosion rates with depth can be related with confidence to the light availability controlling the distribution of phototrophic microendoliths and of grazers feeding on them. This trend is somewhat enforced by the distribution of polychaete borings (*Caulostrepsis*), which clearly decrease with depth as do the foraminiferal attachment scars below their maximum at 15 to 30 m water depth.
- The prevailing decrease in bioerosion rates with time at all depth stations can be attributed to a decrease in bioerosion rate of microendoliths after initial infestation of the pristine substrate and after establishment of mature ichnocoenoses (in the case of the shallow stations). An additional factor can be seen in the successive shading of phototrophic endoliths by epilithic algae cover in the photic zone.
- The orientation of the substrate is known to influence the settlement intensity of encrusting organisms with accretion rates usually being significantly higher on the downward facing compared to the upward facing side of the substrate. This is owing to (1) preference of darker bottom side, (2) enhanced grazing pressure and mortality of encrusters on the top side, and/or (3) enhanced background sedimentation on the top side. In the present study, this observation could be qualitatively confirmed. Especially serpulimorphs and bryozoans were much more abundant on the bottom side at all depth stations. The assessed carbonate accretion rates are referring exclusively to upward facing substrates and can thus be considered as minimum estimations in the given environment.
- The bathymetric dependency of carbonate accretion rates is not as clearly expressed as in the case of the bioerosion rates. A heterogenous bathymetric pattern is found with a distinct maximum at 1 m water depth (governed by the balanid mass occurrence) and a subordinate peak at 15 m (highest population densities of foraminiferans, Serpulidae and Escharellidae).
- There is no general link of carbonate accretion rates and exposure time besides a significant increase from zero to the highest rates recorded during the experiment at the 1 m station owing to the mass occurrence of balanids. When excluding this depth from the evaluation, a constant to weakly increasing tendency towards longer exposure times can be observed at most depths, possibly as a result of weak competition for space on the generally sparsely colonised substrates and the facilitation of larval recruitment after initial biofilm development.

- In contrast to all previous studies, true 2D test surfaces were applied in the Kosterfjord experiment by sealing off all but the upper side of the micrite blocks with epoxy resin. Through the employment of a gravimetrical quantification method, it was possible to directly obtain weight values removed per surface and exposure time, expressed as $g/m^2/y$. This approach avoids the various interpolations required, when applying the widely used various image-analysis methods of sectioned 3D test blocks, as there are (1) interpolation of the surface area and calculated weight losses via substrate densities and calculated removed volume, (2) the decrease in available surface with time, (3) the interference of borings entering the test blocks from different sides, and (4) systematically underestimated bioerosion rates while grazing activity is overestimated owing to the fact that grazing removes substrate, that was already bored by micro- as well as macroborers.

- Comparisons of bioerosion and carbonate accretion rates may be drawn between studies as long as they are not overstressed and when taking methodological pitfalls into account, which are potentially related to gravimetrical versus the image-analysis approach, different substrate compositions, orientations and sizes applied, or the difficulties in separating bioerosion intensities of microborers, macroborers and grazers. Nevertheless, the validity and value of relative trends and patterns observed within one study site, which can in turn be interpreted in relation to other study sites, is unbroken. It must be stressed, however, that experimentally determined bioerosion rates in general are only an approximation for a specific substrate, water depth and exposure time, and balancing over whole reefs or other carbonate sedimentary settings should be undertaken with caution. This is because rates may vary significantly between study sites, settings and even experimental frames due to patchy dispersal and recruitment of bioerosion agents.

- We clearly see a significantly lower pace and rate of bioerosion in the cold-temperate Kosterfjord site if compared with tropical reef settings. The assessed bioerosion rates are to a high degree governed by microbioerosion except for the shallowest stations, where macroborers are additionally relevant. The results are thus best comparable to the results of the Bahamas experiments conducted by Hoskin et al. (1986; 0 to -259 $g/m^2/y$) and Vogel et al. (1996, 2000; 0 to -520 $g/m^2/y$), which applied similar substrates and determined specific microbioerosion rates along extended bathymetric transects. The degree of microbioerosion at the tropical Bahamas site can be judged as roughly 2 to 10 times as high if compared to the Kosterfjord setting, where most rates do not exceed -50 g/m^2/y.

- Overall bioerosion rates in tropical settings generally far exceed the Kosterfjord rates owing to a much higher macroborer and especially grazer activity governing the degree of bioerosion. As a consequence, total bioerosion rates often exceed -1,000 g/m^2/y and the highest overall bioerosion rate reported in the literature amounts to an impressive -25,400 g/m^2/y. Hence, overall bioerosion rates in the tropical reef settings can be judged to exceed the cold-temperate Kosterfjord values by 10 to 100 times.

8.5 Ecological and palaeoenvironmental implications

- When applying the established index ichnocoenoses scheme to the present material, the general trends in bathymetric distribution of the various trace maker groups and their corresponding traces can clearly be confirmed, whereas the individual index ichnocoenoses were of limited applicability. The key-ichnospecies *Fascichnus acinosus* (shallow-euphotic zone II) is too rare in the present material and the rhodophyte trace '*Palaeoconchocelis starmachii*' (shallow-euphotic zone III and deep-euphotic zone) is missing altogether. As suitable substitutes, *Cavernula pediculata* and '*Fascichnus*-form 1', respectively, are suggested. The dysphotic ichnocoenosis is clearly indicated by *Ichnoreticulina elegans* being the only trace produced by a phototrophic organism.

- The biogeographic distribution pattern of phototrophic endoliths reflects a general trend of depletion with higher latitude, as a result of the seasonal fluctuation of light availability and significant differences concerning their lower temperature resistance limit. Consequently, forms that are missing in cold-temperate and/or polar waters bear a value as palaeotemperature and -latitude indicators. In this respect, the record of a number of taxa in the Kosterfjord experiment is extending their known biogeographic distribution northward, as for instance *Eurygonum nodosum* and its trace maker *Mastigocoleus testarum* or *Fascichnus* cf. *acinosus*, produced by *Hyella balani*. The dasycladaceen trace *Fascichnus grandis* does not occur in the present material and was also found to be absent in other North Atlantic sites. The ichnotaxa *Rhopalia catenata* and *Eurygonum pennaforme* are found up to cold-temperate waters and cosmopolitan ichnotaxa are represented for instance by *Cavernula pediculata, Ichnoreticulina elegans, Fascichnus dactylus* and *Scolecia filosa*.

- Endolithic fungi bear a considerable potential as palaeotemperature indicators, but are poorly studied in this respect yet. While the traces *Orthogonum lineare*, *Saccomorpha clava* and '*Flagrichnus*-form 1' have a cosmopolitan distribution and are consequently of no indicative value, both *Saccomorpha terminalis* and '*Flagrichnus*-form 2' potentially indicate cool- to cold-water conditions by their presence. They are of especial value as opposed to palaeotemperature interpretations based on the absence of certain taxa, which are founded on negative evidence and are thus of limited reliability.

- The strong fluctuations of salinity (down to 8 psu) and temperature (0 to 20°C) at the shallow-water stations indicate most phototrophs encountered to be considerably euryhaline and eurytherm, respectively. Most fungal endoliths on the other hand only appear from 15 m downwards reflecting their preference of more stable marine hydrographic conditions or less competition with phototrophs. The fungus *Ostracoblabe implexa* (ichnotaxon: *Orthogonum fusiferum*), however, was almost exclusively encountered at the lagoonal setting and in the *Littorina* shells, confirming its known ability to tolerate reduced salinities and its preference of temperatures above 15°C. It is also known that its growth is significantly reduced under illuminated conditions, explaining its absence in the Iceland spar and the deeply penetrating habit of the trace in the examined shells. Consequently, a deeply boring habit and the appearance of this trace together with cyanobacterial and chlorophyte traces may provide circumstantial evidence for a fossil shallow-water, temperate setting.

- The finding of bathyal communities in comparatively shallow waters, linked to factors that force deeper oceanic water masses to surface, reveals a major potential pitfall in the palaeobathymetric interpretation of fossil *Lophelia* occurrences, which tend to be interpreted as bathyal palaeoenvironments. Strikingly, almost all known exposed ancient *Lophelia* locations derive from tectonically active regions with steep bathymetric gradients and a specific confined topography which could have forced deep water to the near surface.

- The 2 to 10 times lower microbioerosion and 10 to 100 times lower overall bioerosion rates if compared to tropical reef settings, strongly suggest a significant latitudinal gradient. For this decline, the lower ambient temperatures combined with restricted light availability slowing down (bio)chemical processes, as well as specific characteristics in the composition of the bioeroding fauna and flora are held responsible.

The latter differences concern (1) the microendolithic borer spectrum with a considerable depletion towards higher latitudes, (2) the macroborers with important groups (e.g., sipunculans) or taxa (e.g., *Lithophaga*) lacking in high-latitude settings, and most important (3) the significantly lower proportion of grazing bioerosion primarily due to an almost complete lack of effective herbivorous fish, such as parrotfish.

- The Kosterfjord experiment demonstrates, that in the present setting physicochemical dissolution is negligible for shells exposed on the sea floor down to 85 m water depth, clearly identifying bioerosion as the main process of carbonate degradation. Also, early diagenetic precipitation of calcite spar within abandoned microborings (which is common in low-latitudes) was not observed. This holds true not only for the experimental substrates deployed for two years, but also for the taphonomically advanced background sample material investigated. Owing to the long residence of skeletal carbonates exposed on the seafloor as a result of low sedimentation rates, bioerosion – even though at a significantly slower pace than in the Tropics and subtropics – is a strong taphonomic force in high-latitude seas, underlining the comparatively low overall preservation potential of cold-water carbonates.

9 Outlook

In the present volume, the importance of bioerosion as major carbonate degrading process not only in the well-studied tropical reef settings but also in the cold-temperate to polar high latitudes is emphasised. The diversity, both in terms of taxa as well as different ecologic strategies of the various bioerosion agents is outlined, and their applicability as (palaeo-) environmental indicators in several respects is discussed. Also, a first detailed quantification of bioerosion rates in a non-tropical setting is undertaken, pinpointing major latitudinal differences. However, this compilation is only a momentary "snapshot" and much more work is worthwhile to be undertaken to further our knowledge of non-tropical carbonate degradation. Especially promising (not only with respect to non-tropical bioerosion) are investigations leading in the following directions:

- Further experimental studies aiming for a detailed quantification of bioerosion and carbonate accretion in warm-temperate to polar waters are required in order to allow proper balancing of carbon(ate) fluxes in these settings. Such investigations should preferentially utilise extended bathymetrical gradients and could additionally target intertidal to supratidal conditions.

- In future experiments, the relative contribution of grazers, macroborers and microborers should be addressed but circumnavigating the methodological problems arising from the image-analysis methods applied during earlier studies by employing refined gravimetrical methods. Also, the processes and rates of bioerosion *below* the sediment/water interface are poorly understood and are worthwhile tackling, but attention has to be paid to the interference with physicochemical dissolution in this case.

- Extended long-term experiments are required to evaluate and quantify the development of mature ichnocoenoses with respect to macroboring agents, which take much longer to establish compared to tropical reef settings.

- A still wide open field at the interface of biology and (actuo)palaeontology is the study of microendoliths (biotaxa) and their traces (ichnotaxa), many new ones of which are certainly still to be discovered especially in the poorly studied polar waters. While the cyanobacteria and chlorophytes are comparatively well understood, the rhodophytes for instance are barely touched upon and among the marine fungi many riddles remain unsolved as well. Last but not least, bacteria (aside cyanobacteria) are only very rudimentary tackled as yet owing to their minute size rendering them inaccessible to the cast-embedding technique and requiring molecular biological methods such as fluorescence in situ hybridisation (FISH) and others.

- In order to allow neoichnologists to readily compare Recent with fossil borings, and in order of being able to establish ichnotaxa also on the basis of suited Recent holotype material, modifications to the current ICZN are much desired and most sensible in the case of lithic borings, since they morphologically resemble their fossil counterparts in all respects.
- A systematically furthered evaluation of the biogeographic distribution and linked ecological demands of the various bioerosion agents and especially the microendoliths, both in the field and in the laboratory, will push forward the applicability of these organisms and their corresponding traces as (palaeo)environmental indicators, especially in terms of palaeotemperature and -latitude. In parallel, this applicability needs to be proven in palaeoenvironmental reconstructions of fossil carbonates of any age, as it has successfully been done already with the bathymetric index-ichnocoenoses scheme. The latter only requires minor adjustments for the application in cold-temperate to polar settings as it has been indicated by the present study.

References

Abbott RT (1974) American seashells. Van Nostrand Reinhold, New York, 663 pp

Aitken AE, Risk MJ (1988) Biotic interactions revealed by macroborings in arctic bivalve molluscs. Lethaia 21:339-350

Akpan EB (1984) Significance of algal borings and *Acmaea* grazing traces on molluscan shells from Ardyne (west Scotland) postglacial sediments. Palaeogeogr Palaeoclimatol Palaeoecol 47:233-244

Akpan EB (1986) Depth distribution of endolithic algae from the Firth of Clyde: implications for delineation and subdivision of the photic zone. J Mar Biol Assoc UK 66:269-275

Akpan EB, Farrow GE (1984a) Shell-boring algae on the Scottish continental shelf: identification, distribution, bathymetric zonation. Trans R Soc Edinburgh 75: 1-12

Akpan EB, Farrow GE (1984b) Depth of deposition of Early Holocene raised sediments at Irvine deduced from algal borings in mollusc shells. Scott J Geol 20:237-247

Akpan EB, Farrow GE (1985) Shell bioerosion in high-latitude low-energy environments: Firths of Clyde and Lorne, Scotland. Mar Geol 67:139-150

Alderman DJ, Gareth Jones EB (1967) Shell disease of *Ostrea edulis* L. Nature 216:797-798

Alderman DJ, Gareth Jones EB (1971) Physiological requirements of two marine Phycomycetes, *Althornia crouchii* and *Ostracoblabe implexa*. Trans Brit Mycol Soc 57:213-225

Alexander SP, Delaca TE (1987) Feeding adaptations of the foraminiferan *Cibicides refulgens* living epizoically and parasitically on the Antarctic scallop *Admussium colbecki*. Biol Bull 173:136-159

Alexandersson ET (1972) Micritization of carbonate particles: Processes of precipitation and dissolution in modern shallow-marine sediments. Bull Geol Inst Univ Uppsala N S 3:201-236

Alexandersson ET (1974a) Carbonate cementation in coralline algal nodules in the Skagerrak, North Sea: biochemical precipitation in undersaturated waters. J Sed Petrol 44:7-26

Alexandersson ET (1974b) Carbonate cementation in Recent coralline algal constructions. In: Flügel E (ed) Fossil Algae: Recent results and developments. Springer, Berlin Heidelberg, pp 261-269

Alexandersson ET (1976) Actual and anticipated petrographic effects of carbonate undersaturation in shallow seawater. Nature 262:653-657

Alexandersson ET (1977) Carbonate cementation in Recent coralline algal constructions. Flügel E (ed) Fossil Algae: Recent results and developments. Springer, Berlin Heidelberg, 261-269.

Alexandersson ET (1978) Destructive diagenesis of carbonate sediments in the eastern Skagerrak, North Sea. Geology 6:324-327

Alexandersson ET (1979) Marine maceration of skeletal carbonates in the Skagerrak, North Sea. Sedimentology 26:845-852

Al-Thukair AA, Golubic S (1991) Five new *Hyella* species from the Arabian Gulf. In: Hickel B, Anagnostidis K, Komarek J (eds) Cyanophyta /Cyanobacteria – morphology, taxonomy, ecology. Arch Hydrobiol, Suppl 92, Algolog Stud 64:167-197

Al-Thukair AA, Golubic S, Rosen G (1994) New endolithic cyanobacteria from the Bahama Bank and the Arabian Gulf: *Hyella racemus* sp. nov. J Phycol 30: 764-769

Alve E (1995) Benthic foraminiferal responses to estuarine pollution: A review. J Foram Res 25:190-203

Alve E (1996) Benthic foraminiferal evidence of environmental changes in the Skagerrak over the past six decades. Nor Geol Unders Bull 430:85-93

Alve E (1999) Colonisation of new habitats by benthic foraminifera: a review. Earth-Sci Rev 46:167-185

Alve E, Murray JW (1995) Benthic foraminiferal distribution and abundance changes in Skagerrak surface sediments: 1937 (Höglund) and 1992/1993 data compared. Mar Micropal 25:269-288

Alve E, Murray JW (1997) High benthic fertility and taphonomy of foraminifera: a case study of the Skagerrak, North Sea. Mar Micropal 31:157-175

Alve E, Murray JW (1999) Marginal marine environments of the Skagerrak and Kattegat: a baseline study of living (stained) benthic foraminiferal ecology. Palaeogeogr Palaeoclimatol Palaeoecol 146:171-193

Alve E, Nagy J (1986) Estuarine foraminiferal distribution in Sandebukta, a branch of the Oslo Fjord. J Foram Res 16:261-284

Alve E, Olsgard F (1999) Benthic foraminiferal colonisation in experiments with copper contaminated sediments. J Foram Res 29:186-195

Ankel WE (1936) Die Fraßspuren von *Helicon* und *Littorina* und die Funktion der Radula. Zool Anz Suppl 9:174-182

Ankel WE (1937) Wie frisst *Littorina*? Senckenbergiana 19:317-333

Ansell AD (1970) Boring and burrowing mechanisms in *Petricola pholadiformis* Lamarck. J Exp Mar Bio Ecol 4:211-220

Austin WEN, Sejrup HP (1994) Recent shallow water benthic foraminifera from western Norway: ecology and palaeoecological significance. Cushman Found, Spec Publ 32:103-125

Banner FT (1971) A new genus of the Planorbulinidae, an endoparasite of another foraminifer. Rev Españ Micropal 3:113-128

Barnes H, Topinka JA (1969) Effect of the nature of the substratum and the force required to detach a common littoral alga. Am Zool 9:753-758

Barrier P, Di Geronimo I, Montenat C, Roux M, Zibrowius H (1989) Présence de faunes bathyales atlantiques dans le Pliocène et le Pleistocène de Méditerranée (Détroit de Messine, Italie). Bull Soc Géol France 8:787-796

Bathurst RGC (1966) Boring algae, micrite envelopes and lithification of molluscan biosparites. Geol J 5:15-32

Bathurst RGC (1971) Carbonate sediments and their diagenesis. Elsevier, Amsterdam, 620 pp
Batters EAL (1902) A catalogue of the British marine algae. J Bot Suppl 40: 1-107
Bennett MR, Doyle P, Mather AE (1996) Dropstones: their origin and significance. Palaeogeogr Palaeoclimatol Palaeoecol 121:331-339
Bergsten H, Nordberg K, Malmgren B (1996) Recent benthic foraminifera as tracers of water masses along a transect in the Skagerrak, north-eastern North Sea. J Sea Res 35:111-121
Berndt W (1903) Zur Biologie und Anatomie von *Alcippi lampas* Hankock. Z Wiss Zool, A 74:396-457
Bertram MA, Cowen JP (1999) Temporal variations in the deep-water colonisation rates of small benthic foraminifera: the results of an experiment on Cross Seamount. Deep-Sea Res I 46:1021-1049
Beuck L, Freiwald A (2005) Bioerosion patterns in a deep-water *Lophelia pertusa* (Scleractinia) thicket (Propeller Mound, northern Porcupine Seabight). In: Freiwald A, Roberts JM (eds) Cold-water corals and ecosystems. Springer, Berlin Heidelberg, pp 915-936
Blake JA (1969) Systematics and ecology of shell-boring polychaetes from New England. Am Zool 9:813-820
Blake JA, Evans JW (1973) *Polydora* and related genera as borers in mollusc shells and other calcareous substrates. Veliger 15:235-249
Blegvad H (1925) Continued studies on the quantity of fish food in the sea bottom. Rep Danish Mar Biol Stat 30:29-65
Blegvad H (1930) Quantitative investigations of bottom invertebrates in the Kattegat with special reference to the plaice food. Rep Danish Mar Biol Stat 36: 1-55
Boekschoten GJ (1966) Shell borings of sessile epibiontic organisms as palaeoecological guides (with examples from the Dutch coast). Palaeogeogr Palaeoclimatol Palaeoecol 2:333-379
Boekschoten GJ (1970) On bryozoan borings from the Danian at Fakse, Denmark. In: Crimes TP, Harper JC (eds) Trace fossils. Geol J, Spec Iss 3:43-48
Boerboom CM, Smith JE, Risk MJ (1998) Bioerosion and micritization in the deep sea coral *Desmophyllum cristagalli*. Hist Biol 13:53-60
Borg F (1940) On the genus *Tubiporella* and a new boring bryozoan. Zool Bidrag Uppsala 18:415-438
Bornet E, Flahault C (1888) Note sur deux nouveaux genres d'algues perforantes. J Bot, Morot 2:161-165
Bornet E, Flahault C (1889) Sur quelques plantes vivant dans le test calcaire des mollusques. Bull Sci Bot France 36:147-179.
Bosc LAG (1802) Histoire naturelle des vers, contenant leur description et leurs mœurs, avec figures dessinées d'après nature. Deterville, Paris, pp 1-258
Bouchet P, Warén A (1993) Revision of the Northeast Atlantic bathyal and abyssal Mesogastropoda. Soc Ital Malacol, Boll Malacol, Suppl 3:577-840

Brady HB (1879) Notes on some of the reticularian Rhizopoda of the „Challenger" Expedition I. – On new or little known arenaceous types. Quart J Microscop Sci, N S 19:20-63

Bram JB, Page HM, Dugan JE (2005) Spatial and temporal variability in early successional patterns of an invertebrate assemblage at an offshore oil platform. J Exp Mar Biol Ecol 317:223-237

Brattström H (1936) *Ulophysema öresundense* n. gen. et sp., eine neue Art der Ordnung Cirripedia Ascothoracica. Ark Zool 28A:1-10

Brattström H (1937) On the genus *Ulophysema* Brattström with description of a new species from East Greenland. Medd Grønland 115:1-24

Brock RE (1979) An experimental study on the effects of grazing by parrotfishes and role of refuges in benthic community structure. Mar Biol 51:381-388

Broecker WS (2003) The oceanic $CaCO_3$ cycle. In: Holland HD, Turekian KK (eds) Treatise on geochemistry. Elsevier, Amsterdam, pp 529-549

Bromley RG (1970) Borings as trace fossils and *Entobia cretacea* Portlock, as an example. In: Crimes TP, Harper JC (eds) Trace fossils. Geol J, Spec Iss 3: 49-90

Bromley RG (1975) Comparative analysis of fossil and recent echinoid bioerosion. Palaeontology 18:725-739

Bromley RG (1978) Bioerosion of Bermuda reefs. Palaeogeogr Palaeoclimatol Palaeoecol 23:169-197

Bromley RG (1981) Concepts in ichnotaxonomy illustrated by small round holes in shells. Acta Geol Hisp 16:55-64

Bromley RG (1993) Predation habits of octopus past and present and a new ichnospecies, *Oichnus ovalis*. Bull Geol Soc Denmark 40:167-173

Bromley RG (1996) Trace fossils: biology, taphonomy and applications. Chapman & Hall, London, 361 pp

Bromley RG (2004) A stratigraphy of marine bioerosion. In: McIlroy D (ed) The application of ichnology to palaeoenvironmental and stratigraphic analysis. Geol Soc London Spec Publ 228:455-481

Bromley RG (2005) Preliminary study of bioerosion in the deep-water coral *Lophelia*, Pleistocene, Rhodes, Greece. In: Freiwald A, Roberts JM (eds) Cold-water corals and ecosystems. Springer, Berlin Heidelberg, pp 895-914

Bromley RG, Asgaard U (1999) Marble test-blocks for rate of marine bioerosion: the continuing story. In: Clemmensen LB (ed) Abstracts. 19th Europ Meet Int Assoc Sed, Copenhagen, p 43

Bromley RG, Asgaard U (2004) Archaeological distaster a bioerosion goldmine: test-blocks retrieved 2100 years after deposition, Antikythira, Greece. In: Mikuláš R (ed) Abstracts. 4th Int Bioerosion Workshop, Prague, pp 5-6

Bromley RG, D'Alessandro A (1983) Bioerosion in the Pleistocene of Southern Italy: Ichnogenera *Caulostrepsis* and *Maeandropolydora*. Riv Ital Pal Strat 89:283-309

Bromley RG, D'Alessandro A (1984) The ichnogenus *Entobia* from the Miocene, Pliocene and Pleistocene of southern Italy. Riv Ital Pal Strat 90:227-296

Bromley RG, Fürsich FT (1980) Comments on the proposed amendments to the international code of zoological nomenclature regarding ichnotaxa. Z. N. (S.) 1973. Bull Zool Nomencl 37:6-10

Bromley RG, Hanken N-M (1981) Shallow marine bioerosion at Vardø, arctic Norway. Bull Geol Soc Denmark 29:103-109

Bromley RG, Heinberg C (2006) Attachment strategies of organisms on hard substrates: A palaeontological view. Palaeogeogr Palaeoclimatol Palaeoecol 232:429-453

Bromley RG, Martinell J (1991) *Centrichnus,* new ichnogenus for centrically patterned attachment scars on skeletal substrates. Bull Geol Soc Denmark 38:243-252

Bromley RG, Surlyk F (1973) Borings produced by brachiopod pedicles fossil and recent. Lethaia 6:349-365

Bromley RG, Hanken N-M, Asgaard U (1990) Shallow marine bioerosion: preliminary results of an experimental study. Bull Geol Soc Denmark 38: 85-99

Bromley RG, Wisshak M, Glaub I, Boutquelen A (in press) Ichnotaxonomic review of dendriniform borings attributed to foraminifera: *Semidendrina* igen. nov. In: Miller III W (ed) Trace fossils: concepts, problems, prospects. Elsevier, Amsterdam

Bronn HG (1837) Lethaea Geognostica oder Abbildungen und Beschreibungen der für die Gebirge-Formationen bezeichnendsten Versteinerungen. Schweizerbart, Stuttgart, 1:1-544 and plates

Budd DA, Perkins RD (1980) Bathymetric zonation and paleoecological significance of microborings in Puerto Rican shelf and slope sediments. J Sed Petrol 50: 881-904

Bundschuh M, Balog S-J (2000) *Fasciculus rogus* nov. isp., an endolithic trace fossil. Ichnos 7:149-152

Butterfield NJ, Knoll AH, Swett K (1990) A bangiophyte red alga from the Proterozoic of Arctic Canada. Science 250:104-107

Buzas MA (1993) Colonisation rate of foraminifera in the Indian River, Florida. J Foram Res 23:156-161

Buzas MA, Collins LS, Richardson SL, Severin KP (1989) Experiments on predation, substrate preference, and colonisation of benthic foraminifera at the shelfbreak off the Ft. Pierce Inlet, Florida. J Foram Res 19:146-152

Cadee GC (1994) Eider, shelduck, and other predators, the main producers of shell fragments in the Wadden Sea: palaeoecological implications. Palaeontology 37:181-202

Campbell SE (1980) *Palaeoconchocelis starmacchii,* a carbonate boring microfossil from the Upper Silurian of Poland (425 million years old): implications for the evolution of the Bangiaceae (Rhodophyta). Phycologia 19:25-36

Campbell SE, Kazmierczak J, Golubic S (1979) *Palaeoconchocelis starmachii* gen. n., sp. n., an endolithic rhodophyte (Bangiaceae) from the Silurian of Poland. Acta Pal Pol 24:403-408

Cann J-H, Belperio AP, Murray-Wallace CV (2000) Late Quaternary paleosealevels and paleoenvironments inferred from foraminifera, Northern Spencer Gulf, South Australia. J Foram Res 30:29-53

Caram B, Jónsson S (1972) Nouvelle inventaire des algues marines de l'Islande. Acta Bot Islandica 1:5-31

Carefoot T (1977) Pacific Seashores. Douglas, Vancouver, 208 pp

Carriker MR, Van Zandt D (1972) Predatory behaviour of a shell-boring muricid gastropod. In: Winn HE, Oll BL (eds) Behavior of marine animals, 1. Plenum Press, New York, pp 157-244

Carriker MR, Yochelson EL (1968) Recent gastropod boreholes and Ordovician cylindrical borings. Geol Surv Prof Pap 593-B:1-26

Carriker MR, Scott DB, Martin NM (1963) Demineralization mechanism of boring gastropods. In: Sognnaes RF (ed) Mechanisms of hard tissue destruction. Am Assoc Ad Sci 75:55-89

Carter RM (1968) On the biology and palaeontology of some predators of bivalved Mollusca. Palaeogeogr Palaeoclimatol Palaeoecol 4:29-65

Cavaliere AR, Alberte RS (1970) Fungi in animal shell fragments. J Elisha Mitchell Sci Soc 86:203-206

Cedhagen T (1994) Taxonomy and biology of *Hyrrokkin sarcophaga* gen. et sp. n., a parasitic foraminiferan (Rosalinidae). Sarsia 79:65-82

Cerrano C, Bavestrello G, Calcinai B, Cattaneo-Vietti R, Chiantore M, Guidetti M, Sarà A (2001) Bioerosive processes in Antarctic seas. Polar Biol 24:790-792

Cha J-H, Bhaud M (2000) A new experimental approach to assess settlement conditions in tube-building polychaetes: biological implications. Oceanol Acta 23:443- 452

Chazottes V (1996) Étude expérimentale de la bioérosion et de la sédimentogénèse en milieu récifal, effets de l'eutrophisation (Île de la Réunion, Océan Indien). Océanographie (Géologie Marine) 323:787-794

Chazottes V, Le Campion-Alsumard T, Peyrot-Clausade M (1995) Bioerosion rates on coral reefs: interactions between macroborers, microborers and grazers (Moorea, French Polynesia). Palaeogeogr Palaeoclimatol Palaeoecol 113:189-198

Chazottes V, Le Campion-Alsumard T, Peyrot-Clausade M, Cuet P (2002) The effects of eutrophication-related alterations to coral reef communities on agents and rates of bioerosion (Réunion Island, Indian Ocean). Coral Reefs 21: 375-390

Cherchi A, Schroeder R (1991) Perforations branchues dues à des Foraminifères cryptibiotiques dans des coquilles actuelles et fossiles. C R Acad Sci Paris 312:111-115

Chung S-N, Lee K, Feely RA, Sabine CL, Millero FJ, Wanninkhof R, Bullister JL, Key RM, Peng T-H (2003) Calcium carbonate budget in the Atlantic Ocean based on water column inorganic carbon chemistry. Glob Biogeochem Cycl 17:1-15

Clapp WF, Kenk R (1963) Marine borers: an annotated bibliography. Office Naval Res, Dept Navy ACR-74: 1136 pp

Clarke JM (1908) The beginnings of dependent life. New York State Mus Bull 121:146-169

Clokie JJP, Boney AD (1980) *Conchocelis* distribution in the Firth of Clyde: estimates of the lower limits of the photic zone. J Exp Mar Biol Ecol 46: 111-125

Clokie JJP, Boney AD, Farrow GE (1979) The significance of *Conchocelis* as an indicator organism, data from the Firth of Clyde and N.W. Shelf. Brit Phycol J 14:120-121

Clokie JJP, Scoffin TP, Boney AD (1981) Depth maxima of *Conchocelis* and *Phymatolithon rugulosum* on the N.W. Shelf and Rockall Plateau. Mar Ecol Progr Ser 4:131-133

Conway E, Cole K (1977) Studies in the Bangiaceae: structure and reproduction of the conchocelis of *Porphyra* and *Bangia* in culture (Bangiales, Rhodophyceae). Phycologia 16:205-216

Corliss BH, Van Weering TCE (1993) Living (stained) benthic foraminifera within surficial sediments of the Skagerrak. Mar Geol 111:323-335

Crisp DJ (1974) Factors influencing the settlement of marine invertebrate larvae. In: Grant PT, Mackie AM (eds) Chemoreception in marine organisms. Academic Press, New York, pp 177-265

Cutler AH, Flessa KW (1995) Bioerosion, dissolution and precipitation as taphonomic agents at high and low latitudes. Senckenbergiana maritima 25:115-121

Dahl GE (1978) On the existence of a deep countercurrent to the Norwegian Coastal Current in the Skagerrak. Tellus 30:552-556

Davies PJ, Hutchings PA (1983) Initial colonization, erosion and accretion on coral substrates. Experimental results, Lizard Island, Great Barrier Reef. Coral Reefs 2:27-35

Dawes B (1931) Growth and maintenance in the plaice (*Pleuronectes platessa* L.). J Mar Biol Assoc UK 17:877-947

De Bary A (1876) Researches into the nature of the potato-fungus *Phytophthora infestans*. J R Agric Soc 12:239-268

Defrance JLM (1825) Dictionnaire des sciences naturelles dans lequel on traite méthodiquement des différents êtres de la nature, considérés soit en eux-mêmes, d'après l'état actuel de nos connaissances etc. Vol 35:1-534 (NIL-OJO), Levrault, Paris

Delaca TE, Lipps JH (1972) The mechanisms and adaptive significance of attachment and substrate pitting in the foraminiferan *Rosalina globularis* D'Orbigny. J Foram Res 2:68-72

Dewar JM (1908) Notes on the oystercatcher (*Haematopus ostralegus*) with reference to its habit of feeding upon the mussel (*Mytilus edulis*). Zoologica 4:201-212

Dons C (1944) Norges korallrev. Det Kongl Norsk Vidensk Selsk Forh 16:37-82

Drinnan RE (1957) The winter feeding of the oystereater (*Haematopus ostralegus*) on the edible cockl (*Cardium edule*). J Animal Ecol 26:441-469

Ekman T (1896) Beiträge zur Kenntnis des Stieles der Brachiopoden. Z Wiss Zool 62:169-249

Emery KO (1963) Organic transportation of marine sediments. In: Hill MN (ed) The Sea. Interscience, New York, 3, pp 776-793

Emery KO, Tschudy RH (1941) Transportation of rock by kelp. Bull Geol Soc Am 52:855-862

Ercegovic A (1927) Tri nova roda litofiskih cijanoiceja sa jadranske obale. Acta Bot 2:78-84

Ercegovic A (1929) Sur quelques nouveaux types des Cyanophycées lithophytes de la côte Adriatique. Arch Protistenk 66:164-174

Ercegovic A (1932) Ekoloske i socioloske studije o litofitskim cijanoficejama sa Jugoslavenske obale Jadrana. Bull Int Acad Yougoslave Sci Math Nat 26: 129-220

Evans JW (1969) Borers in the shell of the sea scallop *Placopecten magellanicus*. Am Zool 9:775-782

Farran C (1851) *Thalassema neptuni*. Ann Mag Nat Hist 7:156

Farrow GE (1974) On the ecology and sedimentation of the *Cardium* shell-sands and transgressive shellbanks of Traigh Mhor, Island of Barra, Outer Hebrides. Trans R Soc Edinburgh 69:203-227

Farrow GE, Clokie J (1979) Molluscan grazing of sublittoral algal-bored shells and the production of carbonate mud in the Firth of Clyde, Scotland. Trans R Soc Edinburgh 70:139-148

Farrow GE, Fyfe JA (1988) Bioerosion and carbonate mud production on high-latitude shelves. Sed Geol 60:281-297

Farrow GE, Cucci M, Scoffin TP (1978) Calcareous sediments on the nearshore continental shelf of western Scotland. Proc R Soc Edinburgh 76B:55-75

Farrow GE, Allen NH, Akpan EB (1984) Bioclastic carbonate sedimentation on a high-latitude, tide-dominated shelf: northeast Orkney Islands, Scotland. J Sed Petrol 54:373-393

Fischer P (1866) Étude sur les bryozoaires perforants de la famille des Térébriporides. Mus Nat Hist Nat, Nouv Arch 2:293-313

Försterra G, Beuck L, Häussermann V, Freiwald A (2005) Shallow-water *Desmophyllum dianthus* (Scleractinia) from Chile: characteristics of the bioeroding community, heterotrophic interactions and (palaeo)-bathymetrical implications. In: Freiwald A, Roberts JM (eds) Cold-water corals and ecosystems. Springer, Berlin Heidelberg, pp 937-977

Fosså JH, Mortensen PB, Furevik DM (2000) *Lophelia*-korallrev langs norskekysten forekomst og tilstand. Fisken og Havet 2:1-94

Fosså JH, Mortensen, PB Furevik DM (2002) The deep-water coral *Lophelia pertusa* in Norwegian waters: distribution and fishery impacts. Hydrobiologia 471:1-12

Freiwald A (1995) Bacteria-induced carbonate degradation: a taphonomic case study of *Cibicides lobatulus* from a high-boreal carbonate setting. Palaios 10:337-346

Freiwald A (1998) Microbial maceration and carbonate dissolution on cold-temperate shelves. Hist Biol 13:27-35

Freiwald A, Schönfeld J (1996) Substrate pitting and boring pattern of *Hyrrokkin sarcophaga* Cedhagen, 1994 (Foraminifera) in a modern deep-water coral reef mound. Mar Micropal 28:199-207

Freiwald A, Wilson JB (1998) Taphonomy of modern deep, cold-temperate water coral reefs. Hist Biol 13:37-52

Freiwald A, Henrich R, Pätzold J (1997) Anatomy of a deep-water coral reef mound from Stjernsund, West-Finnmark, northern Norway. SEPM Spec Publ 56: 141-161

Friedman GM, Gebelein CD, Sanders JE (1971) Micritic envelopes of carbonate grains are not exclusively of photosynthetic algal origin. Sedimentology 16:89-96

Fujita K (2004) A field colonisation experiment on small-scale distributions of algal symbiont-bearing larger foraminifera on reef rubble. J Foram Res 34:169-179

Garcia-Pichel F (2006) Plausible mechanisms for the boring on carbonates by microbial phototrophs. Sediment Geol 185:205-213

Gaspard D (1989) Quelques aspects de la biodégradation des coquilles de brachiopodes; consequences sur leur fossilisation. Bull Soc Géol France 8:1207-1216

Geitler L (1932) Cyanophyceae. In: Rabenhorst L (ed) Kryptogamen-Flora von Deutschland, Österreich und der Schweiz. Akad Verlagsges, Leipzig, 1196 pp

Gektidis M (1997a) Microbioerosion in Bahamian ooids. Courier Forschinst Senckenberg 201:109-121

Gektidis M (1997b) Vorkommen, Ökologie und Taxonomie von Mikrobohr-organismen in ausgewählten Riffbereichen um die Insel Lee Stocking Island (Bahamas) und One Tree Island (Australien). PhD thesis, Univ Frankfurt, 276 pp

Gektidis M (1999) Development of microbial euendolithic communities: the influence of light and time. Bull Geol Soc Denmark 45:147-150

Glaub I (1994) Mikrobohrspuren in ausgewählten Ablagerungsräumen des europäischen Jura und der Unterkreide (Klassifikation und Palökologie). Courier Forschinst Senckenberg 174:1-324

Glaub I (1999) Paleobathymetric reconstructions and fossil microborings. Bull Geol Soc Denmark 45:143-146

Glaub I (2004) Recent and sub-recent microborings from the upwelling area off Mauritania (West Africa) and their implications for palaeoecology. In: McIlroy D (ed) The Application of Ichnology to Palaeoenvironmental and Stratigraphic Analysis. Geol Soc London, Spec Publ 228:63-77

Glaub I, Bundschuh M (1997) Comparative study on Silurian and Jurassic/Lower Cretaceous microborings. Courier Forschinst Senckenberg 201:123-135

Glaub I, Schmidt H (1994) Traces of endolithic microboring organisms in Triassic and Jurassic bioherms. Kaupia 4:103-112

Glaub I, Vogel K (2004) The stratigraphic record of microborings. Fossils and Strata 51:126-135

Glaub I, Gektidis M, Vogel K (2002) Microborings from different North Atlantic shelf areas – variability of the euphotic zone extension and implications for paleodepth reconstructions. Courier Forschinst Senckenberg 237:25-37

Goldstein S, Belsky M (1964) Axenic culture studies of a new marine phycomycete possessing an unusual type of asexual reproduction. Am J Bot 51: 72-78

Golubic S (1969) Distribution, taxonomy, and boring patterns of marine endolithic algae. Am Zool 9:747-751

Golubic S, Schneider J (1979) Carbonate dissolution. In: Trudinger PA, Swaine DJ (eds) Biogeochemical cycling of mineral-forming elements. Stud Environ Sci 3:107-129

Golubic S, Brent G, LeCampion T (1970) Scanning electron microscopy of endolithic algae and fungi using a multipurpose casting-embedding technique. Lethaia 3:203-209

Golubic S, Perkins RD, Lukas KJ (1975) Boring microorganisms and microborings in carbonate substrates. In: Frey RW (ed) The study of trace fossils. Springer, Berlin Heidelberg, pp 229-259

Golubic S, Krumbein W, Schneider J (1979) The carbon cycle. In: Trudinger PA, Swaine DJ (eds) Biogeochemical cycling of mineral-forming elements. Stud Environ Sci 3:29-45

Golubic S, Friedmann I, Schneider J (1981) The lithobiontic ecological niche, with special reference to microorganisms. J Sed Petrol 51:475-478

Golubic S, Campbell S, Spaeth C (1983) Kunstharzausgüsse fossiler Mikroben-Bohrgänge. Präparator 29:197-200

Golubic S, Radtke G, Le Campion-Alsumard T (2005) Endolithic fungi in marine ecosystems. Trends Microbiol 13:229-235

Grant RE (1826) Notice of a new zoophyte (*Cliona celata* gr.) from the Firth of Forth. Edinburgh New Phil J 1:78-81

Gross O (2000) Influence of temperature, oxygen and food availability on the migrational activity of bathyal benthic foraminifera: Evidence by microcosm experiments. Hydrobiologia 426:123-137

Günther A (1990) Distribution and bathymetric zonation of shell-boring endoliths in recent reef and shelf environments: Cozumel, Yucatan (Mexico). Facies 22:233-262

Gunatilaka A (1976) Thallophyte boring and micritization within skeletal sands from Connemara, western Ireland. J Sed Petrol 46:548-554

Gunkel F (1997) Effects of different substrate roughness on sessile invertebrate recruitment off Lee Stocking Island, Exuma Cays, Bahamas – preliminary results. Proc 8th Int Coral Reef Symp, Panama City, pp 1-5

Gustafsson M, Nordberg K (1999) Benthic foraminifera and their response to hydrography, periodic hypoxic conditions and primary production in the Koljö fjord on the Swedish west coast. J Sea Res 41:163-178

Häntzschel W (1975) Trace fossils and problematica. In: Teichert C (ed) Treatise on Invertebrate Paleontology, Pt W, Miscellanea, Suppl 1. Geol Soc Am, Boulder, Univ Kansas, Lawrence, 269 pp

Hallock P (1988) The role of nutrient availability in bioerosion: consequences to carbonate buildups. Palaeogeogr Palaeoclimatol Palaeoecol 63:275-291

Hancock DA, Urquhart AE (1965) The determination of natural mortality and its causes in an unexploited population of cockles (*Cardium edule* L.). GB Min Agr Fish Food Invest, Ser 2, 24:1-40

Hanken N-M, Bromley RG, Miller J (1996) Plio-Pleistocene sedimentation in coastal grabens, north-east Rhodes, Greece. Geol J 31:393-418

Harriott VJ, Fisk DA (1987) A comparison of settlement plate types for experiments on the recruitment of scleractinian corals. Mar Ecol Progr Ser 37:201-208

Harrison RJ, King JE (1965) Marine mammals. Hutchinson, London, 92 pp

Hassan M (1998) Modification of carbonate substrate by bioerosion and bioaccretion on coral reefs of the Red Sea. Shaker, Aachen, 126 pp

Hein FJ, Risk MJ (1975) Bioerosion of coral heads: inner patch reef, Florida reef tract. Bull Mar Sci 25:133-138

Henderson CM, Styan WB (1982) Description and ecology of Recent endolithic biota from the Gulf Islands and banks in the Straight of Juan de Fuca, British Columbia. Can J Earth Sci 19:1382-1394

Highsmith RC (1980) Geographic patterns of coral bioerosion: a productivity hypothesis. J Exp Biol Ecol 146:177-196

Highsmith RC (1981) Coral bioerosion at Enewetak: agents and dynamics. Int Rev Hydrobiol 66:335-375

Höglund H (1947) Foraminifera in the Gullmarfjord and the Skagerrak. Zool Bidrag Uppsala 26:1-328

Höhnk W (1969) Über den pilzlichen Befall kalkiger Hartteile von Meerestieren. Ber Dt Wiss Komm 20:129-140

Hoeksema BW (1983) Excavation pattern and spiculae dimensions of boring sponge *Cliona caelata* from the SW Netherlands. Senckenbergiana maritima 15: 55-85

Hook JE, Golubic S (1993) Microbial shell destruction in deep-sea mussels, Florida Escarpment. Mar Ecol 14:81-89

Horn MH (1989) Biology of marine herbivorous fishes. Ecology 49:682-694

Hoskin CM, Reed JK, Mook DH (1986) Production and off-bank transport of carbonate sediment, Black Rock, southwest Little Bahama Bank. Mar Geol 73:125-144

Hunt AS, Corliss BH (1993) Distribution and microhabitats of living (stained) benthic foraminifera from the Canadian Arctic Archipelago. Mar Micropal 20: 321-345

Hunter R (1949) The structure and behaviour of *Hiatella gallicana* (Lamarck) and *H. artctica* (L.). Proc R Soc Edinburgh 63:271-289

Hunter R (1977) Sediment production by *Diadema antillarum* on a Barbados fringing reef. Proc 3rd Int Coral Reef Symp, Miami, pp 105-109

Hutchings PA (1981) Polychaete recruitment onto dead coral substrates at Lizard Island, Great Barrier Reef, Australia. Bull Mar Sci 31:410-423

Hutchings PA (1984) A preliminary report on the spatial and temporal patterns of polychaete recruitment on the Great Barrier Reef. In: Hutchings PA (ed) Proc 1st Int Polychaete Conf, Linnean Soc New South Wales, pp 227-237

Hutchings PA (1985) Variability in polychaete recruitment at Lizard Island, Great Barrier Reef: a long-term study and an analysis of its potential impact on coral reef ecosystems. Proc 5th Int Coral Reef Symp, Tahiti, pp 245-250

Hutchings PA, Bamber L (1985) Variability of bioerosion rates at Lizard Island, Great Barrier Reef: preliminary attempts to explain these rates and their significance. Proc 5th Int Coral Reef Symp, Tahiti, pp 333-338

Hutchings PA, Murray A (1982) Patterns of recruitment of polychaetes to coral substrates at Lizard Island, Great Barrier Reef – an experimental approach. Austral J Mar Freshw Res 33:1029-1037

Hutchings PA, Peyrot-Clausade M (2002) The distribution and abundance of boring species of polychaetes and sipunculans in coral substrates in French Polynesia. J Exp Mar Biol Ecol 269:101-121

Hutchings PA, Kiene WE, Reid A, Cunningham RB, Donneley C (1992) Experimental investigations of patterns in the distribution and extend of non-colonial boring communities at Lizard Island, Great Barrier Reef. Coral Reefs 11:23-31

Hutchings P, Peyrot-Clausade M, Osnorno A (2005) Influence of land runoff on rates and agents of bioersion of coral substrates. Mar Poll Bull 51:438-447

Hyman HS (1955) The invertebrates, 4. Echinodermata. McGraw-Hill, New York, 763 pp

ICBN (2000) International Code of Botanical Nomenclature (Saint Louis Code). 16th Int Bot Congr St. Louis, Missouri 1999, Koelz Sci Books, Königstein, 474 pp

ICZN (1999) International Code of Zoological Nomenclature. Nat Hist Mus, London, 306 pp

Jehu TJ (1918) Rock-boring organisms as agents in coast erosion. Scott Geogr Mag 34:1-11

Jones B, Pemberton SG (1987) The role of fungi in the diagenetic alteration of spar calcite. Can J Earth Sci 24:903-914

Jonsson LG, Nilsson PG, Floruta F, Lundälv T (2004) Distributional patterns of macro- and megafauna associated with a reef of the cold-water coral *Lophelia pertusa* on the Swedish west coast. Mar Ecol Progr Ser 284:163-171

Jüch PJW, Boekschoten GJ (1980) Trace fossils and grazing traces produced by *Littorina* and *Lepidochitona*, Dutch wadden sea. Geol Mijnbouw 59:33-42

Kelletat D (1986) Die Bedeutung biogener Formung im Felslittoral Nord-Schottlands. Essener Geogr Arb 14:1-83

Kelletat D (1987) Das Spektrum biogener Formungsprozesse an den Küsten Neuseelands. Berliner Geogr Stud 25:421-454

Kelletat D (1988) Quantitative investigations on coastal bioerosion in higher latitudes: an example from northern Scotland. Geoökodynamik 9:41-51

Kelly SRA, Bromley RG (1984) Ichnological nomenclature of clavate borings. Palaeontology 27:793-807

Kiene WE (1985) Biological destruction of experimental coral substrates at Lizard Island (Great Barrier Reef, Australia). Proc 5th Int Coral Reef Symp, Tahiti, pp 339-344

Kiene WE (1988) A model of bioerosion on the Great Barrier Reef. Proc 6th Int Coral Reef Symp, Townsville, pp 449-454

Kiene WE (1997) Enriched nutrients and their impact on bioerosion: results from ENCORE. Proc 8th Int Coral Reef Symp, Panama City, pp 897-902

Kiene WE, Hutchings PA (1994a) Long-term bioerosion of experimental coral substrates from Lizard Island, Great Barrier Reef. Proc 7th Int Coral Reef Symp, Guam, pp 397-403

Kiene WE, Hutchings PA (1994b) Bioerosion experiments at Lizard Island, Great Barrier Reef. Coral Reefs 13:91-98

Kiene WE, Radtke G, Gektidis M, Golubic S, Vogel K (1995) Factors controlling the distribution of microborers in Bahamian reef environments. In: Schuhmacher H, Kiene WE, Dullo W (coord) Factors controlling Holocene reef growth: an interdisciplinary approach. Facies 32:174-188

Kitazato H (1995) Recolonisation by deep-sea benthic foraminifera: possible substrate preferences. Mar Micropal 26:65-74

Klitgaard-Kristensen D, Buhl-Mortensen L (1999) Benthic foraminifera along an offshore-fjord gradient: a comparison with amphipods and molluscs. J Nat Hist 33:317-350

Kobluk DR, Kahle CF (1978) Geologic significance of boring and cavity-dwelling marine algae. Bull Canad Petroleum Geol 26:362-379

Kobluk DR, Kozelj M (1985) Recognition of a relationship between depth and macroboring distribution in growth framework reef cavities, Bonaire, Netherlands Antilles. Bull Canad Petroleum Geol 33:462-470

Kobluk DR, Risk MJ (1977a) Micritization and carbonate-grain binding by endolithic algae. Bull Am Assoc Petroleum Geol 61:1069-1982

Kobluk DR, Risk MR (1977b) Rate and nature of infestation of a carbonate substratum by a boring alga. J Exp Mar Biol Ecol 27:107-115

Kohlmeyer J (1969) The role of marine fungi in the penetration of calcareous substances. Am Zool 9:741-746

Kornmann P (1959) Die heterogene Gattung *Gomontia*. I. Der sporangiale Anteil, *Codiolum polyrhizum*. Helgoland Wiss Meeresunters 6:229-238

Kornmann P (1960) Die heterogene Gattung *Gomontia*. II. Der fädige Anteil, *Eugomontia sacculata* nov. gen. nov. spec. Helgoland Wiss Meeresunters 7: 59-71

Kornmann P (1961) Über *Codiolum* und *Urospora*. Helgoland Wiss Meeresunters 8:42-57

Kornmann P, Sahling P-H (1980) *Ostreobium quekettii* (Codiales, Chlorophyta). Helgoland Wiss Meeresunters 34:115-122

Krumbein WE, Van der Pers JNC (1974) Diving investigations on biodetoriation by sea-urchins in the rocky sublittoral of Helgoland. Helgoland Wiss Meeresunters 26:1-17

Krutschinna J (1997) Untersuchungen an der Tiefwasserkoralle *Lophelia pertusa* in verschiedenen prä- und postmortalen Stadien unter besonderer Berücksichtigung des mikroendolithischen Befalls. Unpubl masters thesis, Univ Frankfurt, 84 pp

Kylin H (1935) Über einige kalkbohrende Chlorophyceen. K Fysiograf Sällsk Förhandl 5:1-19

Lagerheim G (1885) *Codiolum polyrhizum* n. sp. Ett bidrag till kännedomen om slätget *Codiolum*. K Vetenskaps Akad Förhandl 42:21-31

Lagerheim G (1886) Notes sur le *Mastigocoleus*, nouveau genre des algues marines de l'ordre des Phycochromacées. Notarisia 1:65-69

Lambers P, Boekschoten GJ (1986) On fossil and recent borings produced by acrothoracic cirripeds. Geol Mijnbouw 65:257-268

Langer MR (1993) Epiphytic foraminifera. Mar Micropal 20:235-265

Larkum AWD, Steven ADL (1994) ENCORE: The effect of nutrient enrichment on coral reefs. 1. Experimental design and research programme. Mar Poll Bull 29:112-120

Le Bris S, Le Campion-Alsumard T, Romano JC (1998) Caractéristiques du feutrage algal des récifs coralliens de Polynésie française soumis à différentes intensités de bioérosion. Oceanol Acta 21:695-708

Le Campion-Alsumard T (1975) Etude expérimentale de la colonisation d'éclats de calcite par les Cyanophycées endolithes marines. Cah Biol Mar 16:177-185

Le Campion-Alsumard T (1978) Les Cyanophycées endolithes marines. Systématique, ultrastructure, écologie et biodestruction. PhD thesis, Univ Aix-Marseille II, vol 1, 198 pp

Le Campion-Alsumard T (1979) Les Cyanophycées endolithes marines. Systématique, ultrastructure, écologie et biodestruction. Oceanol Acta 2: 143-156

Le Campion-Alsumard T, Golubic S (1985) *Hyella caespitosa* Bornet et Flahault and *Hyella balani* Lehmann (Pleurocapsales, Cyanophyta): a comparative study. Arch Hydrobiol Suppl 71:119-184

Le Campion-Alsumard T, Romano J-C, Peyrot-Clausade M, Le Campion J, Paul R (1993) Influence of some coral reef communities on the calcium carbonate budget of Tiahure Reef (Moorea, French Polynesia). Mar Biol 115:685-693

Le Campion-Alsumard T, Golubic S, Hutchins P (1995a) Microbial endoliths in skeletons of live and dead corals: *Porites lobata* (Moorea, French Polynesia). Mar Ecol Progr Ser 117:149-157

Le Campion-Alsumard T, Golubic S, Priess K (1995b) Fungi in corals: symbiosis or disease? Interaction between polyps and fungi causes pearl-like skeleton biomineralization. Mar Ecol Progr Ser 117:137-147

Lehmann E (1903) Über *Hyella balani* nov. spec. Nytt Mag Natruvid 41:77-87

Liebau A (1984) Grundlagen der Ökobathymetrie. In: Luterbacher H (ed) Paläobathymetrie. Pal Kursb 2:149-184

Lindberg B, Mienert J (2005) Postglacial production by cold-water corals on the Norwegian Shelf and their role in the global carbonate budget. Geology 33: 537-540

Linke P, Lutze GF (1993) Microhabitat preferences of benthic foraminifera – a static concept or a dynamic adaptation to optimise food acquisition? Mar Micropal 20:215-234

Littler MM, Littler DS, Blair SM, Norris JN (1985) Deepest known plant life discovered on an uncharted seamount. Science 227:57-59

Lönöy N (1954) A comparative study of *Phoronis ovalis* Wright from Norwegian, Swedish, and Brazilian waters. Univ Bergen Naturv Rekke 1953:1-29

Lüning K (1985) Meeresbotanik: Verbreitung, Ökophysiologie und Nutzung der marinen Makroalgen. Thieme, Stuttgart, 375 pp

Lukas KJ (1978) Depth distribution and form among common microboring algae from the Florida continental shelf. Geol Soc Am, Abstr with Progr 10:448

Lukas KJ, Golubic S (1981) New endolithic cyanophytes from the North Atlantic Ocean: I. *Cyanosaccus piriformis* gen. et sp. nov. J Phycol 17:224-229

Lukas KJ, Golubic S (1983) New endolithic cyanophytes from the North Atlantic Ocean. II. *Hyella gigas* Lukas & Golubic sp. nov. from the Florida Continental Margin. J Phycol 19:129-136

Lund S (1959) The marine algae of East Greenland. I. Taxonomic part. Medd Grønland 156:1-247

Lundälv T, Jonsson LG (2000) Inventering av Koster – Väderöområdet med ROVteknik. En pilotstudie. Swedish EPA Rep 5079:1-75

Lundälv T, Jonsson LG (2003) Mapping of deep-water corals and fishery impacts in the north-east Skagerrak, using acoustical and ROV survey techniques. Proc 6th Underwater Sci Symp, London, pp 23-26

Mackenzie FT (1998) Our changing planet: an introduction to earth system science and global environmental change. Prentice-Hall, Upper Saddler River, 486 pp

Magwood (1992) Ichnotaxonomy: a burrow by any other name…? In: Maples CG, West RR (eds) Trace Fossils. Pal Soc Am, Knoxville, pp 15-33

Maughan BC (2001) The effects of sedimentation and light on recruitment and development of a temperate, subtidal, epifaunal community. J Exp Mar Biol Ecol 256:59-71

May JA, Perkins RD (1979) Endolithic infestation of carbonate substrates below the sediment-water interface. J Sed Petrol 49:357-378

Mayoral E (1988) *Pennatichnus* nov. icnogen.; *Pinaceocladichnus* nov. icnogen. e *Iramena.* huellas de bioerosion debidas a Bryozoa perforantes (Ctenostomata, Plioceno Inferior) en la Cuenca del Bajo Guadalquivir. Rev Españ Pal 3: 13-22

McLean RF (1967) Measurement of beach rock erosion by some tropical marine gastropods. Bull Mar Sci 17:551-561

Menge BA, Lubchenco J (1981) Community organization in temperate and tropical rocky intertidal habitats: prey refuges in relation to consumer pressure gradients. Ecol Monogr 51:429-450

Mikkelsen N, Erlenkeuser H, Killingley JS, Berger WH (1982) Norwegian corals: radiocarbon and stable isotopes in *Lophelia pertusa*. Boreas 11:163-171

Milligan HN (1916) Observations on the feeding habits of the purple-tipped sea-urchin. Zoologist 20:81-99

Milliman JD (1993) Production and accumulation of calcium carbonate in the ocean: budget of a nonsteady state. Glob Biogeochem Cycl 7:927-957

Milliman JD, Troy PJ, Balch WM, Adams AK, Li Y-H, Mackenzie FT (1999) Biologically mediated dissolution of calcium carbonate above the chemical lysocline? Deep-Sea Res I 46:1653-1669

Moen FE, Svenson E (2004) Marine fish & invertebrates of Northern Europe. KOM, Kristiansund, 608 pp

Moodley L, Troelstra SR, Van Weering TCE (1993) Benthic foraminiferal response to environmental change in the Skagerrak, northeastern North Sea. Sarsia 78:129-139

Moore CH, Shedd WW (1977) Effective rates of sponge bioerosion as a function of carbonate production. Proc 3rd Int Coral Reef Symp, Miami, pp 499-505

Morton JE, Miller MC (1968) The New Zealand Sea Shore. Collins, London Auckland, 638 pp

Morton JE (2004) Seashore ecology of New Zealand and the Pacific. Bateman, Auckland, 504 pp

Moses CA (2000) Field rock block exposure trials. Z Geomorph N F, Suppl 120: 33-50

Müller OF (1776) Zoologiae Danicae Prodromus, seu Animalium Daniae et Norvegiae Indigenarum Characteres, Nomina, et Synonyma Imprimis Popularium. Havniae, pp 1-282

Müller E, Löffler W (1992) Mykologie. Thieme, Stuttgart, 367 pp

Mullineaux LS, Delaca TE (1984) Distribution of Antarctic benthic foraminifers settling on the pecten *Adamussium colbecki*. Polar Biol 3:185-189

Mullineaux LS, Mills SW, Goldman E (1998) Recruitment variation during a pilot colonisation study of hydrothermal vents (9°50'N, East Pacific Rise). Deep-Sea Res II 45:441-464

Murray JW (1991) Ecology and distribution of benthic foraminifera. In: Lee JJ, Anderson OR (eds) Biology of Foraminifera. Academic Press, London, pp 221-253

Murray JW, Alve E (1999) Natural dissolution of modern shallow water benthic foraminifera: taphonomic effects on the palaeoecological record. Palaeogeogr Palaeoclimatol Palaeoecol 146:195-209

Naylor LA, Viles HA (2002) A new technique for evaluating short-term rates of coastal bioerosion and bioprotection. Geomorphology 47:31-44

Neumann AC (1966) Observations on coastal erosion in Bermuda and measurements of the boring rate of the sponge *Cliona lampa*. Limnol Oceanogr 11:92-108

Newman WA, Ross A (1971) Antarctic Cirripedia. Antarctic Res Ser 14:1-246

Nielsen KSS, Nielsen JK, Bromley RG (2003) Palaeoecological and ichnological significance of microborings in Quartenary foraminifera. Pal Electr 6:13 pp

Nielsen R (1972) A study of the shell-boring marine algae around the Danish island Læsø. Bot Tidsskr 67:245-269

Nielsen R (1980) A comparative study of five marine Chaetophoracea. Brit Phycol J 15:131-138

Nielsen R (1987) Marine algae within calcareous shells from New Zealand. New Zealand J Bot 25:425-438

Nielsen R, Correa J (1987) A comparative study of *Gomontia polyrhiza* and *Chlorojackia pachyclados* gen. et sp. nov. (Chlorophyta). Can J Bot 65: 2467-2472

Nielsen R, Gunnarsson K (2001) Seaweeds of the Faroe Islands. An annotated checklist. Fródskaparrit 49:45-108

Nielsen R, Kristiansen A, Mathiesen L, Mathiesen H (1995) Distributional index of the benthic marine macroalgae of the Baltic Sea area. Acta Bot Fenn 155: 1-70

Noble JPA, Logan A, Webb GR (1976) The Recent *Terebratulina* community in the rocky subtidal zone of the Bay of Fundy, Canada. Lethaia 9:1-17

Norton SF (1988) Role of the gastropod shell and operculum in inhabiting predation by fishes. Science 241:92-94

Olempska E (1986) Endolithic microorganisms in Ordovician ostracod valves. Acta Pal Pol 31:229-236

Orbigny AD d' (1826) Tableau méthodique de la classe des Céphalopodes: Ann Sci Nat Paris, Ser I 7:245-314

Orton JH (1926) On the rate of growth of *Cardium edule*, 1. Experimental observations. J Mar Biol Assoc UK 14:239-280

Osorno A, Peyrot-Clausade M, Hutchings PA (2005) Patterns and rates of erosion in dead *Porites* across the Great Barrier Reef (Australia) after 2 and 4 years of exposure. Coral Reefs 24:292-303

Otter GW (1932) Rock-burrowing echinoids. Biol Rev 7:89-107

Pari N, Peyrot-Clausade M, Le Campion-Alsumard T, Hutchings PA, Chazottes V, Golubic S, Le Campion J, Fontaine MF (1998) Bioerosion of experimental substrates on high islands and on atoll lagoons (French Polynesia) after two years of exposure. Mar Ecol Progr Ser 166:110-130

Pari N, Peyrot-Clausade M, Hutchings PA (2002) Bioerosion of experimental substrates on high islands and atoll lagoons (French Polynesia) during 5 years of exposure. J Exp Mar Biol Ecol 276:109-127

Parker FL (1953) In: Parker FL, Phleger FB, Pierson JF (eds) Ecology of foraminifera from San Antonio Bay and environs, Southwest Texas. Cushman Found Foram Res Spec Publ 2:1-75 [note: reference to an untitled essay]

Parker WK, Jones TR (1860) On the nomenclature of foraminifera. IV. The species enumerated by Lamarck. Ann Mag Nat Hist 3:29-40

Paull CK, Hecker B, Commeau R, Freeman-Lynde RP, Neumann C, Corso WP, Golubic S, Hook JE, Sikes E, Curray J (1984) Biological communities at the Florida escarpment resemble hydrothermal vent taxa. Science 226:965-967

Pech D, Ardisson P-L, Bourget E (2002) Settlement of a tropical marine epibenthic assemblage on artificial panels: influence of substratum heterogeneity and complexity scales. Estuar Coast Shelf Sci 55:743-750

Pedersen PM (1976) Marine benthic algae from southernmost Greenland. Medd Grønland 199:1-80

Perkins RD, Tsentas CI (1976) Microbial infestations of carbonate substrates planted on the St. Croix shelf. Geol Soc Am Bull 87:1615-1628

Perry CT, Macdonald IA (2002) Impacts of light penetration on the bathymetry of reef microboring communities: implications for the development of microendolithic trace assemblages. Palaeogeogr Palaeoclimatol Palaeoecol 189:101-113

Peyrot-Clausade M, Hutchings PA, Richards G (1992) The distribution and successional patterns of macroborers in marine *Porites* at different stages of degradation on the barrier reef, Tiahura, Moorea, French Polynesia. Coral Reefs 11:161-166

Peyrot-Clausade M, Le Campion-Alsumard T, Harmelin-Vivien M, Romano J-C, Chazottes V, Pari N, Le Campion J (1995a) La bioérosion dans le cycle des carbonates: essais de quantification des precessus en Polynésie francaise. Bull Soc Geol France 166:85-94

Peyrot-Clausade M, Le Campion-Alsumard T, Hutchings PA, Le Campion J, Payri C, Fontaine MF (1995b) Initial bioerosion and bioaccretion on experimental substrates in high island and atoll lagoons (French Polynesia). Oceanol Acta 18:531-541

Peyrot-Clausade M, Chazottes V, Pari N (1999) Bioerosion in the carbonate budget of two Indo-Pacific reefs: La Réunion (Indian Ocean) and Mooréa (Pacific Ocean). Bull Geol Soc Denmark 45:151-155

Pia J (1937) Die kalklösenden Thallophyten. Arch Hydrobiol 31:264-328, 341-398

Pickerill RK (1994) Nomenclature and taxonomy of invertebrate trace fossils. In: Donovan SK (ed) The Palaeobiology of Trace Fossils. Wiley, Chichester, pp 3-42

Plewes CR, Palmer TJ, Haynes JR (1993) A boring foraminiferan from the Upper Jurassic of England and Northern France. J Micropal 12:83-89

Pohowsky RA (1974) Notes on the study and nomenclature of boring Bryozoa. J Sed Petrol 48:556-564

Pohowsky RA (1978) The boring ctenostomate Bryozoa: taxonomy and paleobiology based on cavities in calcareous substrata. Bull Am Pal 73:1-192

Porter D, Lingle WL (1992) Endolithic thraustochytrid marine fungi from planted shell fragments. Mycologia 84:289-299

Purchon RD (1955) The structure and function of the British Pholadidae (rock-boring Lamellibranchia). Proc Zool Soc London 124: 859-887

Radtke G (1991) Die mikroendolithischen Spurenfossilien im Alt-Tertiär West-Europas und ihre palökologische Bedeutung. Courier Forschinst Senckenberg 138:1-185

Radtke G (1993) The distribution of microborings in molluscan shells from Recent reef environments at Lee Stocking Island, Bahamas. Facies 29:81-92

Radtke G, Golubic S (2005) Microborings in mollusc shells, Bay of Safaga, Egypt: morphometry and ichnology. Facies 51:118-134

Radtke G, Hofmann K, Golubic S (1997) A bibliographic overview of micro- and macroscopic bioerosion. Courier Forschinst Senckenberg 201:307-340

Radwański A (1977) Present-day types of trace in the Neogene sequence; their problems of nomenclature and preservation. In: Crimes TP, Harper JC (eds) Trace Fossils 2. Geol J, Spec Iss 9:227-264

Rao CP (1996) Modern carbonates – tropical, temperate, polar. Carbonates, Univ Tasmania, 206 pp

Reaka-Kudla ML, Feingold JS, Glynn W (1996) Experimental studies of rapid bioerosion of coral reefs in the Galápagos Islands. Coral Reefs 15:101-107

Ribes T, Salvado H, Romero J, Del Pilar Gracia M (2000) Foraminiferal colonisation on artificial seagrass leaves. J Foram Res 33:192-201

Ridgwell A, Zeebe RE (2005) The role of the global carbonate cycle in the regulation and evolution of the earth system. Earth Planet Sci Lett 234:299-315

Rindsberg AK (1990) Ichnological consequences of the 1985 International Code of Zoological Nomenclature. Ichnos 1:59-63

Rippka R (1972) Photoheterotrophy and chemoheterotrophy among unicellular blue-green algae. Arch Microbiol 87:93-98

Rogers AD (1999) The biology of *Lophelia pertusa* (Linnaeus 1758) and other deep-water reef-forming corals and impacts from human activities. Internat Rev Hydrobiol 84:315-406

Rokoengen K, Østmo SR (1985) Shallow geology off Fedje, western Norway. IKU-Rep 24.1459/01/25, 20 pp

Rueness J, Brattegard T, Lein TE, Küfner Lein R, Pedersen A, Sørlie AC (2001) Class Chorophyceae (division Chlorophyta) – green algae (grønnalger). In: Brattegard T, Holthe T (eds) Distribution of marine, benthic macro-organisms in Norway – A tabulated catalogue. Res Rep 2001-3, Direct Nat Managem, Trondheim, pp 6-12

Rützler K (1975) The role of burrowing sponges in bioerosion. Oecologia 19: 203-216

Ruiz GM, Lindberg DR (1989) A fossil record for trematodes: extend and potential uses. Lethaia 22:431-438

Sato-Okoshi W, Okoshi K (1997) Survey of the genera *Polydora*, *Boccardiella* and *Boccardia* (Polychaeta, Spionidae) in Barkley Sound (Vancouver Island, Canada), with special reference to boring activity. Bull Mar Sci 60:482-493

Savazzi E (1999) Boring, nestling and tube-dwelling bivalves. In: Savazzi E (ed) Functional morphology of the invertebrate skeleton. Wiley, Chicester, pp 205-238

Schafer CT, Cole FE, Frobel D, Rice N, Buzas MA (1996) An in situ experiment on temperature sensitivity of nearshore temperate benthic foraminifera. J Foram Res 26:53-63

Scheffer VB (1958) Seals, sea lions and walruses; a review of the pinnipedia. Univ Press, Stanford, 179 pp

Schlichter D, Kampmann H, Conrady S (1997) Trophic and photoecology of endolithic algae living within coral skeletons. Mar Ecol 18:299-317

Schmidt H (1992) Mikrobohrspuren ausgewählter Faziesbereiche der tethyalen und germanischen Trias (Beschreibung, Vergleich und bathymetrische Interpretation). Frankfurter Geowiss Arb A 12:1-228

Schmidt H, Freiwald A (1993) Rezente gesteinsbohrende Kleinorganismen des norwegischen Schelfs. Natur und Museum 123:149-155

Schneider J (1976) Biological and inorganic factors in the destruction of limestone coasts. Contr Sed 6:1-123

Schneider J, Le Campion-Alsumard T (1999) Construction and destruction of carbonates by marine and freshwater cyanobacteria. Eur J Phycol 34: 417-426

Schneider J, Torunski H (1983) Biokarst on limestone coasts, morphogenesis and sediment production. Mar Ecol 4:45-63

Schneider RR, Schulz D, Hensen C (2000) Marine carbonates: their formation and destruction. In: Schulz D, Zabel M (eds) Marine geochemistry. Springer, Berlin Heidelberg, pp 283-307

Schnick H (1992) Zum Vorkommen der Bohrspur *Hyellomorpha microdendritica* Vogel, Golubic & Brett im oberen Obermaastricht Mittelpolens. Z Geol Wiss 20:109-124

Schroeder JH (1972) Calcified filaments of an endolithic algae in Recent Bermuda reefs. N Jb Geol Pal Mh 1:16-33

Schuhmacher H (1977) Initial phases in reef development, studied at artificial reef types off Eilat, (Red Sea). Helgoland Wiss Meeresunters 30:400-411

Scoffin TP, Alexandersson ET, Bowes GE, Clokie JJ, Farrow GE, Milliman JD (1980a) Recent, temperate, sub-photic, carbonate sedimentation: Rockall Bank, northeast Atlantic. J Sed Petrol 50:331-356

Scoffin TP, Stearn CW, Boucher D, Frydl P, Hawkins CM, Hunter IG, MacGeachy JK (1980b) Calcium carbonate budget of a fringing reef on the west coast of Barbados. Bull Mar Sci 30:475-508

Scott PJB (1988) Factors influencing the susceptibility of submarine calcareous substrates to attack by bioeroders. In: Thompson MF, Sarojini R, Nagabhushanam R (eds) Marine biodetoriation: Advances – techniques, applicable to Indian Ocean. IBH Publ, New Delhi, pp 681-691

Scott PJB, Moser KA, Risk MJ (1988) Bioerosion of concrete and limestone by marine organisms: A 23 year experiment from Jamaica. Mar Poll Bull 19: 219-222

Seidenkrantz MS (1993) Subrecent changes in the foraminiferal distribution in the Kattegat and Skagerrak, Scandinavia: anthropogenic influence and natural causes. Boreas 22:383-395

Seilacher A (1969) Paleoecology of boring barnacles. Am Zool 9:705-719

Sergeant WAS (1975) Plant trace fossils. In: Frey RW (ed) The study of trace fossils. Springer, Berlin Heidelberg, pp 163-180

Silén L (1946) On two new groups of Bryozoa living in shells of molluscs. Ark Zool 38:1-7

Silén L (1947) On the anatomy and biology of Penetrantiidae and Immergentiidae (Bryozoa). Ark Zool 40A:1-48

Silén L (1956) On shell-burrowing Bryozoa and Phoronis from New Zealand. Trans R Soc New Zealand 84:93-96

Smith AM (1988) Preliminary steps toward formation of a generalized budget for cold-water carbonates. Sediment Geol 60:323-331

Smith AM, Nelson CS (2003) Effects of early sea-floor processes on the taphonomy of temperate shelf skeletal carbonate deposits. Earth-Sci Rev 63:1-31

Soule JD, Soule DF (1969) Systematics and biogeography of burrowing bryozoans. Am Zool 9:791-802

Squires DF (1964) Fossil coral thickets in Wairarapa, New Zealand. J Paleont 38:904-915

Stearn CW, Scoffin TP (1977) Carbonate budget of a fringing reef, Barbados. Proc 3rd Int Coral Reef Symp, Miami, pp 471-476

Strømgren T (1971) Vertical and horizontal distribution of *Lophelia pertusa* (Linné) in Trondheimsfjorden on the west coast of Norway. K Norske Vidensk Selsk Skr 6:1-9

Svansson A (1975) Physical and chemical oceanography of the Skagerrak and the Kattegat, I. Open sea conditions. Rep Fish Board Swed Inst Mar Res 1:1-88

Tavernier A, Campbell SE, Golubic S (1992) A complex marine shallow-water boring trace: *Dendrorete balani* n. ichnogen. et ichnospec. Lethaia 25:303-310

Taylor PD, Wilson MA (2003) Palaeoecology and evolution of marine hard substrate communities. Earth-Sci Rev 62:1-103

Taylor PD, Wilson MA, Bromley RG (1999) *Leptichnus*, a new ichnogenus for etchings made by cheilostome bryozoans into calcareous substrates. Palaeontology 42:595-604

Taylor WR (1957) Marine algae of the northeastern coast of North America. Univ Michigan Press, Michigan, 509 pp

Terquem O (1875) Essai sur le classement des animaux qui vivent sur la plage et dans les environs de Dunkerque. Paris, pp 1-54

Titschack J, Freiwald A (2005) Growth, deposition, and facies of Pleistocene bathyal coral communities from Rhodes, Greece. In: Freiwald A, Roberts JM (eds) Cold-water corals and ecosystems. Springer, Berlin Heidelberg, pp 41-59

Tomascik T (1991) Settlement patterns of Caribbean scleractinian corals on artificial substrata along an eutrophication gradient, Barbados, West Indies. Mar Ecol Progr Ser 77:261-269

Tomlinson JT (1953) A burrowing barnacle of the genus *Trypetesa* (Order Acrothoracica). J Washington Acad Sci 43:373-381

Tomlinson JT (1969) Shell-burrowing barnacles. Am Zool 9:837-840

Tomlinson JT (1987) The burrowing barnacles. In: Southward AJ (ed) Crustacean Issues 5, Barnacle Biology. Balkema, Rotterdam, pp 63-71

Trewin NH, Welsh W (1976) Formation and composition of graded estuarine shell bed. Palaeogeogr Palaeoclimatol Palaeoecol 19:219-230

Tribollet A, Payri C (2001) Bioerosion of the coralline alga *Hydrolithon onkodes* by microborers in the coral reefs of Moorea, French Polynesia. Oceanol Acta 24:329-342

Tribollet A, Decherf G, Hutchings PA, Peyrot-Clausade M (2002) Large-scale spatial variability in bioerosion of experimental coral substrates on the Great Barrier Reef (Australia): importance of microborers. Coral Reefs 21: 424-432

Trudgill ST (1976) The marine erosion of limestone on Aldabra Atoll, Indian Ocean. Z Geomorph Suppl 26:164-200

Trudgill ST (1983) Measurements of rates of erosion of reefs and reef limestones. In: Barnes DJ (ed) Perspective on coral reefs. Austral Inst Mar Sci Contr 20: 256-262

Trudgill ST (1987) Bioerosion of intertidal limestone, Co. Clare, Eire – 3. Zonation, process and form. Mar Geol 74:111-121

Trudgill ST, Crabtree RW (1987) Bioerosion of intertidal limestone, Co. Clare, Eire – 2: *Hiatella arctica*. Mar Geol 74:99-109

Trudgill ST, Smart PL, Friederich H, Crabtree RW (1987) Bioerosion of intertidal limestone, Co. Clare, Eire – 1: *Paracentrotus lividus*. Mar Geol 74:85-98

Tudhope AW, Risk MJ (1985) Rate of dissolution of carbonate sediments by microboring organisms, Davies Reef, Australia. J Sed Petrol 55:440-447

Van der Pers JNC (1978) Bioerosion by *Polydora* (Polychaeta, Sedentaria, Vermes) off Helgoland, Germany. Geol Mijnbouw 57:465-478

Van Dover CL, Berg CJ, Turner RD (1988) Recruitment of marine invertebrates to hard substrates at deep-sea hydrothermal vents on the East Pacific Rise and Galápagos spreading center. Deep-Sea Res 35:1833-1849

Van Weering TCE, Qvale G (1983) Recent sediments and foraminiferal distribution in the Skagerrak, northeastern North Sea. Mar Geol 52:75-99

Vénec-Peyré M-T (1996) Bioeroding foraminifera: a review. Mar Micropal 28: 19-30

Vogel K (1997) Bioerosion in rezenten Riffbereichen: Experimente vor den Inseln Bahamas und des Großen Barriereriffs. Natur und Museum 127:198-208

Vogel K, Glaub I (2004) 450 Millionen Jahre Beständigkeit in der Evolution endolithischer Mikroorganismen? Sitzber Wiss Ges Univ Frankfurt 42: 1-42

Vogel K, Marincovich LJ (2004) Palaeobathymetric implications of microborings in Tertiary strata of Alaska, USA. Palaeogeogr Palaeoclimatol Palaeoecol 206: 1-20

Vogel K, Golubic S, Brett CE (1987) Endolith associations and their relation to facies distribution in the Middle Devonian of New York State, U.S.A. Lethaia 20:263-290

Vogel K, Bundschuh M, Glaub I, Hofmann K, Radtke G, Schmidt H (1995) Hard substrate ichnocoenoses and their relations to light intensity and marine bathymetry. N Jb Geol Pal Abh 195:49-61

Vogel K, Kiene W, Gektidis M, Radtke G (1996) Scientific results from investigations of microbial borers and bioerosion in reef environments. Göttinger Arb Geol Pal, Sonderb 2:139-143

Vogel K, Balog S-J, Bundschuh M, Gektidis M, Glaub I, Krutschinna J, Radtke G (1999) Bathymetrical studies in fossil reefs, with microendoliths as paleoecological indicators. Profil 16:181-191

Vogel K, Gektidis M, Golubic S, Kiene WE, Radtke G (2000) Experimental studies on microbial bioerosion at Lee Stocking Island, Bahamas and One Tree Island,

Great Barrier Reef, Australia: implications for paleoecological reconstructions. Lethaia 33:190-204
Voigt E (1965) Über parasitische Polychaeten in Kreide-Austern sowie einige andere in Muschelschalen bohrende Würmer. Pal Z 39:193-211
Voigt E (1971) Fremdskulpturen an Steinkernen von Polychaeten-Bohrgängen aus der Maastrichter Tuffkreide. Pal Z 45:144-153
Voigt E (1975) Tunnelbaue rezenter und fossiler Phoronidea. Pal Z 49: 135-167
Voigt E (1977) On grazing traces produced by the radula of fossil and recent gastropods and chitons. In: Crimes TP, Harper JC (eds) Trace fossils. Geol J, Spec Iss 3:335-346
Wahrberg R, Eliason A (1926) Ny local för levande *Lophelia prolifera* (Pallas) vid svensk kust. Fauna och Flora 1926:256-260
Walker G, Jacob E (1798) In: Adams G (ed), Essays on the Microscope. Dillon and Keating, London [Note: this is a reference to an untitled essay]
Waller TR (1980) Scanning electron microscopy of shell and mantle in the order Arcoida (Mollusca: Bivalvia). Smithson Contr Zool 313:1-58
Warme JE (1975) Borings as trace fossils, and the process of marine bioerosion. In: Frey RW (ed) The study of trace fossils. Springer, Berlin Heidelberg, pp 181-227
Wefer G, Lutze G (1978) Carbonate production by benthic foraminifera and accumulation in the western Baltic. Limnol Oceanogr 23:992-996
Wefer G, Richter W (1976) Colonisation of artificial substrates by foraminifera. Kieler Meeresforsch, Sonderh 3:72-75
Wefer G, Balzer W, Bodungen B von, Suess E (1987) Biogenic carbonates in temperate and subtropical environments: production and accumulation, saturation state and stable isotope composition. In: Rumohr J, Walger E, Zeitschel B (eds) Seawater-sediment interactions in coastal waters. Springer, Berlin Heidelberg, pp 263-302
Wilkinson M (1974) Investigations on the autecology of *Eugomontia sacculata* Kornmann, a shell-boring alga. J Exp Mar Biol Ecol 16:19-27
Wilkinson M (1975) The occurence of shell-boring *Phaeophila* species in Britain. Brit Phycol J 10:235-240
Wilkinson M, Burrows EM (1970) *Eugomontia sacculata* Kornmann in Britain and North America. Brit Phycol J 5:235-238
Wilkinson M, Burrows EM (1972) The distribution of marine shell-boring green algae. J Mar Biol Assoc UK 52:59-65
Williamson WC (1858) On the Recent Foraminifera of Great Britain. R Soc Publ, London, 107 pp
Wilson JB (1967) Palaeoecological studies on shell-beds and associated sediments in the Solway Firth. Scott J Geol 3:329-371
Wilson JB (1982) Shelly faunas associated with temperate offshore tidal deposits. In: Stride AH (ed) Offshore tidal sands – processes and deposits. London, pp 126-171

Wilson MA (2005) Marine bioerosion bibliography. http://www.wooster.edu/geology/bioerosion/bioerosion.html
Wisshak M, Porter D (in press) The new ichnogenus *Flagrichnus* – a palaeoenvironmental indicator for cold-water settings? Ichnos
Wisshak M, Rüggeberg A (2006) Colonisation and bioerosion of experimental substrates by benthic foraminiferans from euphotic to aphotic depths (Kosterfjord, SW Sweden). Facies 52:1-17
Wisshak M, Freiwald A, Lundälv T, Gektidis M (2005a) The physical niche of bathyal *Lophelia pertusa* in a non-bathyal setting: environmental controls and palaeoecological implications. In: Freiwald A, Roberts JM (eds) Cold-water corals and ecosystems. Springer, Berlin Heidelberg, pp 979-1001
Wisshak M, Gektidis M, Freiwald A, Lundälv T (2005b) Bioerosion along a bathymetric gradient in a cold-temperate setting (Kosterfjord, SW Sweden): an experimental study. Facies 51:93-117
Woelke CE (1957) The flatworm *Pseudostylochus ostreophagus* Hyman, a predator on oysters. Proc Natl Shellfish Assoc 47:62-67
Wollast R (1994) The relative importance of biomineralization and dissolution of $CaCO_3$ in the global carbon cycle. Bull Inst Océanogr Monaco, Spec Iss 13: 13-35
Wollenburg J (1995) Benthische Foraminiferen als Wassermassen-, Produktions- und Eisdriftanzeiger im Arktischen Ozean. Ber Polarforsch 179:1-227
Wilkinson M (1974) Investigations on the autecology of *Eugomontia sacculata* Kornm., a shell-boring alga. J Exp Mar Biol Ecol 16:19-27
Yonge CM (1963) Rock-boring organisms. In: Sognnaes RF (ed) Mechanisms of hard tissue destruction. Am Assoc Ad Sci 75:1-24
Yonge CM (1964) Rock borers. Sea Frontiers 10:106-116
Young HR, Nelson CS (1985) Biodegradation of temperate-water skeletal carbonates by boring sponges on the Scott Shelf, British Columbia, Canada. Mar Geol 65:33-45
Young HR, Nelson CS (1988) Endolithic biodegradation of cool-water skeletal carbonates on Scott shelf, northwestern Vancouver Island, Can J Sed Geol 60:251-267
Zebrowski G (1937) New genera of Cladochytriaceae. Ann Missouri Bot Gard 23: 553-564
Zeff ML, Perkins RD (1979) Microbial alteration of Bahamian deep-sea carbonates. Sedimentology 26:175-201
Zhang Y, Golubic S (1987) Endolithic microfossils (Cyanophyta) from early Proterozoic stromatolites, Hebei, China. Acta Micropal Sinensis 4:1-12
Zubia M, Peyrot-Clausade M (2001) Internal bioerosion of *Acropora formosa* in Réunion (Indian Ocean): microborer and macroborer activities. Oceanol Acta. 24:251-262

Appendix 1

PVC plates

Water depth [m]	Exposure time [days]	Plates [n]	Serpulidae [counts]	Serpulidae [ind/m²]	Spirorbidae [counts]	Spirorbidae [ind/m²]	Tubuliporidae [counts]	Tubuliporidae [ind/m²]	Escharellidae [counts]	Escharellidae [ind/m²]	Balanidae [counts]	Balanidae [ind/m²]	Antedonidae [counts]	Antedonidae [ind/m²]
1	177	4	0	0	0	0	0	0	0	0	0	0	0	0
7	178	6	16	267	0	0	0	0	0	0	4	67	0	0
15	178	6	66	1,100	0	0	0	0	3	50	1	17	0	0
30	175	6	1	17	0	0	0	0	0	0	0	0	0	0
50	180	6	6	100	3	50	0	0	0	0	0	0	0	0
85	177	6	41	683	28	467	0	133	0	0	0	0	13	217
1	389	6	0	0	0	0	0	0	0	0	702	11,700	0	0
7	392	6	7	117	0	0	0	0	0	0	0	0	0	0
15	392	6	72	1,200	0	0	1	17	2	33	0	0	0	0
30	390	6	77	1,283	0	0	0	0	0	0	11	183	0	0
50	390	6	74	1,233	0	0	0	0	0	0	0	0	0	0
85	390	6	33	550	34	567	13	217	0	0	0	0	6	100
1	lost	0												
7	726	6	3	50	0	0	0	0	1	17	0	0	0	0
15	726	6	3	50	0	0	0	0	5	83	1	17	0	0
30	724	6	4	67	2	17	0	0	0	0	0	0	0	0
50	724	6	2	33	1	17	6	100	1	17	0	0	0	0
85	725	6	2	33	5	83	10	167	0	0	0	0	0	0
lagoon 1	178	6	0	0	0	0	0	0	0	0	4	67	0	0
lagoon 1	513	6	0	0	0	0	0	0	0	0	2	33	0	0
lagoon 1	all times	6	0	0	0	0	0	0	0	0	3	50	0	0
mean 1	all times	3	0	0	0	0	0	0	0	0	351	5,850	0	0
mean 7	all times	6	9	145	0	0	0	0	0	6	1	22	0	0
mean 15	all times	6	47	783	0	0	0	6	3	55	1	11	0	0
mean 30	all times	6	27	456	1	6	0	0	0	0	4	61	0	0
mean 50	all times	6	27	455	1	22	2	33	0	6	0	0	0	0
mean 85	all times	6	25	422	22	372	8	172	0	0	0	0	6	106
all depths	all times	6	19	323	3	57	1	30	1	10	51	856	1	15

Carbonate plates

Water depth [m]	Exposure time [days]	Plates [n]	Serpulidae [counts]	Serpulidae [ind/m²]	Spirorbidae [counts]	Spirorbidae [ind/m²]	Tubuliporidae [counts]	Tubuliporidae [ind/m²]	Escharellidae [counts]	Escharellidae [ind/m²]	Balanidae [counts]	Balanidae [ind/m²]	Antedonidae [counts]	Antedonidae [ind/m²]
1	177	5	0	0	0	0	0	0	0	0	0	0	0	0
7	178	6	22	381	0	0	0	0	0	0	0	0	0	0
15	178	6	76	1,317	0	0	0	0	0	0	3	52	0	0
30	175	6	0	0	0	0	0	0	0	0	0	0	0	0
50	180	6	9	156	0	0	0	0	0	0	0	0	0	0
85	177	6	38	659	23	399	24	416	1	17	0	0	0	0
1	389	6	0	0	0	0	0	0	0	0	881	15,271	0	0
7	392	6	41	711	0	0	0	0	0	0	0	0	0	0
15	392	6	98	1,699	0	0	0	0	2	35	0	0	0	0
30	390	6	90	1,560	1	17	0	0	0	0	7	121	0	0
50	390	6	86	1,491	3	52	0	0	0	0	1	17	0	0
85	390	6	33	572	28	503	14	243	0	0	0	0	0	0
1	lost	0												
7	726	5	23	478	0	0	0	0	1	21	1	21	0	0
15	726	4	13	338	0	0	0	0	1	26	1	26	0	0
30	724	6	13	225	2	35	1	17	0	0	1	17	0	0
50	724	6	13	225	6	104	0	0	0	0	4	69	0	0
85	725	6	5	87	14	243	2	35	0	0	0	0	0	0
lagoon 1	178	6	1	17	0	0	0	0	0	0	1	17	0	0
lagoon 1	513	5	4	83	0	0	0	0	0	0	2	42	0	0
lagoon 1	all times	6	3	50	0	0	0	0	0	0	2	30	0	0
mean 1	all times	4	0	0	0	0	0	0	0	0	441	7,636	0	0
mean 7	all times	6	29	523	0	0	0	0	0	7	0	7	0	0
mean 15	all times	5	62	1,118	0	0	0	0	1	20	1	26	0	0
mean 30	all times	6	34	595	1	17	0	6	0	0	3	46	0	0
mean 50	all times	6	36	624	3	52	0	0	0	0	2	29	0	0
mean 85	all times	6	25	439	22	382	13	231	0	6	0	0	0	0
all depths	all times	5	27	479	4	64	2	34	0	5	64	1,110	0	0

PVC + carbonate plates

Water depth [m]	Exposure time [days]	Plates [n]	Serpulidae [counts]	Serpulidae [ind/m²]	Spirorbidae [counts]	Spirorbidae [ind/m²]	Tubuliporidae [counts]	Tubuliporidae [ind/m²]	Escharellidae [counts]	Escharellidae [ind/m²]	Balanidae [counts]	Balanidae [ind/m²]	Antedonidae [counts]	Antedonidae [ind/m²]
lagoon 1	all times	6	1	25	0	0	0	0	0	0	2	40	0	0
mean 1	all times	4	0	0	0	0	0	0	0	0	396	6,743	0	0
mean 7	all times	6	19	334	0	0	0	0	0	6	1	15	0	0
mean 15	all times	6	55	951	0	0	0	3	2	38	1	19	0	0
mean 30	all times	6	31	525	1	12	0	3	0	0	3	54	0	0
mean 50	all times	6	32	540	2	37	1	17	0	3	1	14	0	0
mean 85	all times	6	25	431	22	377	11	202	0	3	0	0	3	53
all depths	all times	6	23	401	4	61	2	32	0	7	58	983	0	8

Appendix 2

PVC plates

Water depth	Exposure time	Plates	*Cibicides lobatulus*		*Planorbulina mediterranensis*		*Tholosina vesicularis*		*Lituotuba lituiformis*		*Nubecularia lucifuga*		All foraminiferans	
[m]	[days]	[n]	[counts]	[ind/m²]	[counts]	[ind/m²]	[counts]	[ind/m²]	[counts]	[ind/m²]	[counts]	[ind/m²]	[counts]	[ind/m²]
1	177	4	0	0	0	0	0	0	0	0	0	0	0	0
7	178	6	0	0	0	0	0	0	0	0	0	0	0	0
15	178	6	16	267	0	0	2	33	0	0	0	0	18	300
30	175	6	0	0	0	0	0	0	0	0	0	0	0	0
50	180	6	6	100	0	0	0	0	0	0	0	0	6	100
85	177	6	5	83	0	0	0	0	0	0	0	0	5	83
1	389	6	0	0	0	0	0	0	0	0	0	0	0	0
7	392	6	318	5,300	0	0	2	33	403	6,717	11	183	734	12,233
15	392	6	1,205	20,083	9	150	0	0	1,021	17,017	13	217	2,248	37,467
30	390	6	1,174	19,567	8	133	112	1,867	109	1,817	0	0	1,403	23,384
50	390	6	207	3,450	0	0	0	0	4	67	0	0	211	3,517
85	390	6	163	2,717	1	17	0	0	422	7,033	0	0	586	9,767
1	lost	0												
7	726	6	1,095	18,250	398	4,967	7	117	0	0	105	1,750	1,605	25,084
15	726	6	2,965	49,417	20	333	33	550	1,315	21,917	122	2,033	4,455	74,250
30	724	6	84	1,400	34	567	9	150	1	17	2	33	130	2,167
50	724	6	503	8,383	8	133	10	167	25	417	0	0	546	9,100
85	725	6	493	8,217	10	167	0	0	499	8,317	0	0	1,002	16,701
lagoon 1	178	6	1	17	0	0	6	100	0	0	0	0	7	0
lagoon 1	513	6	0	0	0	0	21	350	0	0	0	0	21	0
lagoon 1	all times	6	1	9	0	0	14	225	0	0	0	0	14	0
mean 1	all times	3	0	0	0	0	0	0	0	0	0	0	0	0
mean 7	all times	6	471	7,850	133	1,656	3	50	134	2,239	39	644	780	12,439
mean 15	all times	6	1,395	23,256	10	161	12	194	779	12,978	45	750	2,240	37,339
mean 30	all times	6	419	6,989	14	233	40	672	37	611	1	11	511	8,517
mean 50	all times	6	239	3,978	3	44	3	56	10	161	0	0	254	4,239
mean 85	all times	6	220	3,672	4	61	0	0	307	5,117	0	0	531	8,850
all depths	all times	6	392	6,536	23	308	10	171	181	3,015	12	201	619	10,198

Carbonate plates

Water depth	Exposure time	Plates	*Cibicides lobatulus*		*Planorbulina mediterranensis*		*Tholosina vesicularis*		*Lituotuba lituiformis*		*Nubecularia lucifuga*		All foraminiferans	
[m]	[days]	[n]	[counts]	[ind/m²]	[counts]	[ind/m²]	[counts]	[ind/m²]	[counts]	[ind/m²]	[counts]	[ind/m²]	[counts]	[ind/m²]
1	177	5	0	0	0	0	0	0	0	0	0	0	0	0
7	178	6	0	0	0	0	0	0	0	0	0	0	0	0
15	178	6	0	0	0	0	0	0	0	0	0	0	0	0
30	175	6	0	0	0	0	0	0	0	0	0	0	0	0
50	180	6	0	0	0	0	6	104	0	0	0	0	6	104
85	177	6	1	17	0	0	0	0	0	0	0	0	1	17
1	389	6	0	0	0	0	0	0	0	0	0	0	0	0
7	392	6	6	104	0	0	0	0	73	1,265	14	243	93	1,612
15	392	6	36	624	3	52	0	0	556	9,637	27	468	622	10,781
30	390	6	23	399	6	104	0	0	4	69	0	0	33	572
50	390	6	17	295	0	0	1	17	3	52	0	0	21	364
85	390	6	9	156	0	0	0	0	0	0	0	0	9	156
1	lost	0												
7	726	5	94	1,955	5	104	0	0	488	10,150	743	15,454	1,330	27,663
15	726	4	31	806	4	104	2	52	924	24,128	51	1,326	1,012	26,416
30	724	6	51	884	23	399	0	0	0	0	0	0	74	1,283
50	724	6	39	676	10	173	2	35	15	260	0	0	66	1,144
85	725	6	9	156	2	35	0	0	397	6,881	0	0	408	7,072
lagoon 1	178	6	0	0	0	0	0	0	0	0	0	0	0	0
lagoon 1	513	6	0	0	0	0	0	0	0	0	0	0	0	0
lagoon 1	all times	6	0	0	0	0	0	0	0	0	0	0	0	0
mean 1	all times	4	0	0	0	0	0	0	0	0	0	0	0	0
mean 7	all times	6	33	686	2	35	0	0	187	3,805	252	5,232	474	9,758
mean 15	all times	5	22	477	2	52	1	17	493	11,255	26	598	545	12,399
mean 30	all times	6	25	428	10	168	0	0	1	23	0	0	36	618
mean 50	all times	6	19	324	3	58	3	52	6	104	0	0	31	537
mean 85	all times	6	6	110	1	12	0	0	132	2,294	0	0	139	2,415
all depths	all times	6	15	289	3	46	1	10	117	2,497	40	833	175	3,675

PVC + carbonate plates

Water depth	Exposure time	Plates	*Cibicides lobatulus*		*Planorbulina mediterranensis*		*Tholosina vesicularis*		*Lituotuba lituiformis*		*Nubecularia lucifuga*		All foraminiferans	
[m]	[days]	[n]	[counts]	[ind/m²]	[counts]	[ind/m²]	[counts]	[ind/m²]	[counts]	[ind/m²]	[counts]	[ind/m²]	[counts]	[ind/m²]
lagoon 1	all times	6	0	4	0	0	7	113	0	0	0	0	7	0
mean 1	all times	4	0	0	0	0	0	0	0	0	0	0	0	0
mean 7	all times	6	252	4,268	67	845	2	25	161	3,022	146	2,938	627	11,099
mean 15	all times	6	709	11,866	6	107	6	106	636	12,117	36	674	1,393	24,869
mean 30	all times	6	222	3,708	12	201	20	336	19	317	0	6	273	4,568
mean 50	all times	6	129	2,151	3	51	3	54	8	133	0	0	143	2,388
mean 85	all times	6	113	1,891	2	37	0	0	220	3,705	0	0	335	5,633
all depths	all times	6	204	3,413	13	177	5	90	149	2,756	26	517	397	6,937

Taxonomic index

Bold: Taxon (including 'cf.' and '?') described in the systematic part (Sect. 4)
Bold italics: Taxon (including 'cf.' and '?') depicted

D

E

F

G

H

I

K

L

M

U

V

Printing: Krips bv, Meppel
Binding: Stürtz, Würzburg